CERAMIC PETROGRAPHY and HOPEWELL INTERACTION

A Dan Josselyn Memorial Publication

CERAMIC PETROGRAPHY and HOPEWELL INTERACTION

James B. Stoltman

THE UNIVERSITY OF ALABAMA PRESS
Tuscaloosa

The University of Alabama Press
Tuscaloosa, Alabama 35487-0380
uapress.ua.edu

Typeface: Bembo and Myriad

Manufactured in the United States of America
Cover photograph: Photomicrograph of granite-tempered vessel 40-8 from Icehouse Bottom; courtesy of James B. Stoltman.
Cover design: Todd Lape/Lape Designs

Ebook editions of this title feature color illustrations for figures

1.1, 2.4–2.5, 2.9–2.10, 2.12, 3.3, 5.1, 5.3, 5.5–5.8,
6.1–6.2, 6.4, 7.1, 8.1–8.2, 9.1–9.3, and 10.1.

∞

The paper on which this book is printed meets the minimum requirements of American National Standard for Information Sciences—Permanence of Paper for Printed Library Materials, ANSI Z39.48-1984.

Library of Congress Cataloging-in-Publication Data

Stoltman, James B., 1935–
Ceramic petrography and Hopewell interaction / James B. Stoltman.
pages cm
Includes bibliographical references and index.
ISBN 978-0-8173-1859-8 (cloth : alk. paper) — ISBN 978-0-8173-8807-2 (e book) 1. Hopewell culture. 2. Petrology in archaeology. 3. Indian pottery—Analysis. 4. Ceramics—Analysis. 5. Indians of North America—Antiquities.
I. Title.
E99.H69S76 2015
970.004'97—dc23
2014034310

For Sallie
who was always there along the way

Contents

Illustrations

FIGURES

TABLES

Acknowledgments

This work could not even have been attempted, much less completed, without the help of numerous friends and colleagues. In roughly chronological order (my sample collecting formally began in 1985), I extend my sincerest thanks to the following people, beginning with those based in Ohio institutions: N'omi Greber and David Brose, then at the Cleveland Museum of Natural History; Olaf Prufer and Mark Seeman at Kent State University; Martha Potter Otto of the Ohio Historical Society; Rick Yerkes and Bill Dancey at The Ohio State University. Jim Brown (of Northwestern University) was instrumental in getting sherds from Mound City for thin sectioning and together we collected most of the soil samples recorded in Table 3.15. Sissel Schroeder, in a trip across southern Ohio, collected glacial Lake Tight sediment samples for me. In addition Jim Morton of Columbus, Ohio, was an interested supporter of this work and was instrumental in procuring a sherd for thin sectioning from the Knight Hollow Rockshelter simple-stamped vessel. It was an enormous pleasure to meet and visit with Robert Harness about working on samples from "his" site. I must also offer an additional special thank you to N'omi Greber for her extraordinary cooperation throughout this project, particularly for her role in procuring sherds from the Peabody Museum at Harvard University where the Turner site collections reside.

The Mann site is a critical component of this study, and it was Bret Ruby and Christine Shriner of Indiana University who encouraged and aided my interest in the site. During the course of our discussions, Dr. Shriner offered to send me the thin sections from the site that she had prepared and analyzed, an offer that I most gratefully accepted. Continuing down the Ohio River, it was a pleasure to work with Brian Butler and Mackenzie Caldwell Rohm on the surprising Middle Woodland materials from the 11Mx109 site in southern Illinois. Terrance Martin, Associate Curator at the Illinois State Museum, was most cooperative in providing sherds for thin sectioning from the two "exotic" vessels originally recovered from the Rutherford Mound by Mike Fowler (1957).

Turning now to the Southeast, Vincas Steponaitis, Director of the Research Laboratories of Anthropology at the University of North Carolina, graciously provided me sherds from the important Garden Creek Mound 2. At the University of Tennessee in Knoxville, Jefferson Chapman, Director of the Frank H. McClung Museum, kindly provided sherds for thin sectioning from Icehouse Bottom. From the University of Georgia, my old classmate and friend, David Hally, provided me access to the important Tunacunnhee and Mandeville collections. Jerald Milanich and Ann Cordell of the Anthropology Division of the Florida Museum of Natural History in Gainesville were most cooperative in selecting sherds for this study from their Crystal River collections. Frankie Snow, my collaborator on an earlier study of Swift Creek ceramics, obtained several sherds from Kolomoki for thin sectioning. Robert Mainfort introduced me to Pinson Mounds, then later provided me with sherds from the site for thin sectioning that became the basis for our collaboration on a paper (Stoltman and Mainfort 2002) that contradicted an earlier paper in which he was senior author (Mainfort et al. 1997). A brave fellow of great integrity! It was John Connaway of the Mississippi Department of Archives and History who sent me sherds from nine Marksville Incised vessels from the Yazoo Basin of Mississippi to be used for comparison with the grog-tempered vessels that were surprisingly turning up in Middle Woodland sites to the east. I want to thank Scott Keith for getting me involved in the analysis of the large sample of thin sections from the Leake site. This volume would not be the same without their inclusion!

Several scholars working in Kentucky heard of my research and offered to send me suspected exotic sherds for thin section evaluation. I gratefully acknowledge the receipt of such sherds (all of which proved to be imports to northern Kentucky) from the following: Gwynn Henderson and Richard Jefferies (from the University of Kentucky) who sent sherds from the Bentley, Hansen, and Camargo sites; Richard Stallings and Steve Mocas (from Louisville, Kentucky) who sent a sherd from the incredible Panther Rock vessel; and Michael Richmond (from Lexington, Kentucky) who forward a sherd from the equally incredible Amburgey vessel.

Additionally, three colleagues here at the University of Wisconsin-Madison, Gordon Medaris and Robert Dott of the Department of Geology, and James Burton from my own department, were endless sources of wise counsel. I thank them most sincerely for their willingness to share their time and wisdom with me.

Most of the thin sections were flawlessly prepared by Brian Hess, master craftsman and proprietor of Hess Petrographic. Thanks, Brian.

Finally, two anonymous reviewers were sage and constructive readers of the manuscript. Their observations were taken to heart and the final manuscript benefited enormously from their conscientious efforts.

CERAMIC PETROGRAPHY and HOPEWELL INTERACTION

Introduction

The main goal of this book is to use the technique of petrography to uncover and explain evidence of cultural interaction among Hopewell societies of the Ohio Valley region and contemporary people of the Southeastern United States, as evidenced by the circulation of pottery vessels among sites that bore evidence of participation in the Hopewell Interaction Sphere (Caldwell 1964; Seeman 1979). The way petrography is applied to pottery analysis in this volume is unusual. Some would call it idiosyncratic. I like to think of it as unique, hopefully in a productive way. Traditionally, the petrographic analysis of ceramics has been mainly qualitative in character identifying (with great accuracy) what minerals and rock fragments are present in a ceramic body. In this study, however, several explicitly quantitative measures have been added to traditional petrography in order to provide greater depth and objectivity to the characterization of ceramic materials. It is hoped such data will provide the reader with more explicit and objective bases for evaluating the data and the views being presented.

For purposes of this volume, the Southeast is that region east of the Mississippi River and south of the Ohio River. Participation in the Hopewell Interaction Sphere is traditionally recognized through the presence of various distinctive artifact types and raw materials (e.g., Seeman 1979:304 and 313). Notably, pottery vessels are never included in such lists. It was this realization that contributed to my embarking on this study in the mid-1980s. As I worked hard to retrain myself in the use of a petrographic microscope (I had an undergraduate degree in geology), it occurred to me that with all the interest in goods associated with the Hopewell Interaction Sphere, the compositional analysis of pottery could add a valuable, complementary perspective. So was born the idea of conducting a comparative analysis of pottery vessels recovered from Ohio Hopewell and related sites in the Southeast. My goals were as much methodological as substantive, because I was attempting to develop and apply a methodology that employed quantitative as well as traditional qualitative techniques to the compositional analysis of pottery (e.g., Stoltman 1989 and 1991).

Beginning in 1985, I contacted friends and colleagues who were conversant in things Hopewell. Almost continuously since that time, through the intercession and aid of countless people to whom I am enormously indebted, I was able to amass an incredible collection of thin-sectioned pottery sherds from most of the major Hopewell sites of Ohio and all of the major Southeastern sites for which there is direct archaeological evidence of Hopewell interaction. I am sure there are other sites out there that could be included in the study, but the time has come to end the sample collecting and to report the results. In the ensuing chapters wherein specific sites are identified as possible or even probable sources of "trade" vessels, it should be understood that such claims are offered not as proof but as the most probable alternatives within the framework of the current sample of Middle Woodland sites.

The volume is organized into eleven chapters and this introduction. Chapter 1 provides a brief introduction to petrography, the main method used in this study. In *The Dictionary of Geological Terms* (Bates and Jackson 1984:379), petrography is defined as "That branch of geology dealing with description and systematic classification of rocks . . . by means of microscopic examination of thin sections." The term is often used synonymously with petrology, which is "broader in scope than petrography," dealing also with the origins and history of rocks (Bates and Jackson 1984:380). The term petrography will be used throughout this volume with the understanding that in this application of the technique to the analysis of pottery, the *identification* of minerals and rocks in thin section (not specifically mentioned though clearly implied in the above definition) is the main purpose of the approach.

Chapter 2 is the longest, because it records the compositional analyses of the local ceramics from seven of the major Ohio Hopewell centers, Liberty, Mound City, Hopewell, Seip, Marietta, Tremper, and Turner. In addition a smaller sample from McGraw, a residential hamlet, is included plus suspected exotic sherds from Knight Hollow, Murphy, Newark, and Grimes. The goal of this chapter is to establish the compositional parameters of the locally made pottery from the Hopewell centers of Ohio.

Before detailed analyses were conducted, the ceramic samples from each site were subjected to an initial screening. Based on this screening, vessels that were either stylistically anomalous or possessed nonlocal tempers (especially grog, shell, and sand) were temporarily set aside as "Possibly Nonlocal" in order to allow the establishment of reliable local compositional signatures for Ohio Hopewell pottery based on those vessels for which there was little doubt of their local manufacture. Detailed consideration of the "Possibly

Nonlocal" ceramics was deferred until Chapter 10, after the compositional analysis of ceramics from the major Southeastern sites had been completed so that productive comparisons could be made between Ohio and Southeastern ceramics in search of identifiable "trade" vessels.

Two unexpected findings in the compositional analysis of Ohio Hopewell ceramics are noteworthy. First, the local pottery at Tremper is grog tempered, the first and only such occurrence in southern Ohio. Second, the two main ceramic series in the Hopewell sites of southern Ohio, Scioto, and Hopewell were consistently made with different recipes. The former, presumably primarily for everyday, utilitarian purposes, was tempered with more abundant and coarser rock particles, while the latter, presumably for more specialized purposes, has fewer and finer temper inclusions.

The next two chapters report on the ceramic analyses of three sites in the lower Ohio Valley, the important Mann site in southwestern Indiana (Chapter 3), and two sites in southern Illinois, the Rutherford Mound and the recently reported 11Mx109 site (Chapter 4). Christine Shriner's excellent petrographic analysis of Mann ceramics has been published previously (Ruby and Shriner 2005). The present study complements that analysis by recording grog temper separately from the paste so that the Mann data could be integrated into the database of this report. In addition, not only did Dr. Shriner generously allow me to analyze her ceramic thin sections but she also provided thin sections of several soil samples from in and around the site, which I also analyze in Chapter 3. One of the surprises of the Mann analysis was the discovery that seven grog-tempered sherds recovered from five sites extending from Garden Creek in North Carolina to Pinson Mounds in west Tennessee were compositionally indistinguishable from the local pottery at Mann.

The Rutherford Mound is well known for its wide array of classic Hopewellian grave goods accompanied by a nearly complete simple-stamped vessel with tetrapodal appendages, an obvious Southeastern import (Fowler 1957:27). A sherd from this vessel, along with another from a cord-marked jar (Fowler 1957:Plate 1a) that Prufer (1965:22) had attributed to the Ohio type, McGraw Cordmarked, were obtained from the Illinois State Museum for thin section analysis. Unfortunately, no local pottery was requested (my oversight), so no comparisons with local pottery from the site are possible for these two vessels. Nonetheless, on mineralogical grounds the simple-stamped vessel can definitely be accepted as a Southeastern import while Prufer may well have been correct in suspecting an Ohio source for the cord-marked jar.

11Mx109 (discussed in Chapter 4) is included in this study despite lacking evidence of Hopewell interaction because, as a Late Woodland site with a previously unrecognized Middle Woodland component, its grog-tempered ceramics extend the geographic distribution of grog tempering along the lower Ohio River in Middle Woodland contexts from southern Ohio (Tremper) almost to the Mississippi River. The site is seemingly closely related to Mann, although Mackenzie Caldwell, who reported on the Middle Woodland component of the site for her Master's thesis, sees closer connections with the Mississippi Valley (personal communication).

With Chapters 5, 6, and 7, the focus of attention turns to the ceramics of the Southeast. The sites included were all selected because of the presence of unambiguous evidence of Hopewell interaction. Each of the three chapters represents a different region: the Blue Ridge (Chapter 5), the Appalachian Plateau, (Chapter 6), and the Coastal Plain (Chapter 7). The primary content of these chapters is the compositional makeup of the local ceramics of each of the sites for eventual use in determining the origins of vessels believed to be intrusive to the Ohio Hopewell sites.

Two of the Blue Ridge sites, Garden Creek and Icehouse Bottom, are especially noteworthy for the presence of substantial numbers of rocker-stamped sherds, a classic Ohio Hopewell decorative style. Eight such Hopewell-style vessels from Garden Creek were thin sectioned, and all were determined to be local products. Icehouse Bottom is unique in having the only Hopewell-style vessels (n=2) from a Southeastern site that can be accepted as imports from Ohio. Not only did Ohio pottery reach Icehouse Bottom, but two unmistakably Icehouse Bottom-like vessels were recovered at the Seip site.

Two important sites in north Georgia represent the Appalachian Plateau and Ridge and Valley provinces in this study, Tunacunnhee and Leake (Chapter 6). The petrographic data from Leake provide a most important object lesson, namely, that the common expectation that the local ceramics of a site will be characterized by mineralogical homogeneity is not necessarily true.

Chapter 7 considers three Coastal Plain sites, Mandeville, Kolomoki, and Crystal River, which play prominent roles in any discussion of Hopewell interaction with this region. With minor exceptions, these sites provide a clear-cut compositional signature for the Coastal Plain, that is, quartz-dominated, sand-tempered pottery. This region was surely the major source of marine shell that was so prominently circulated within the Hopewell Interaction Sphere, but the ceramic evidence for contact between the Coastal Plain and Ohio is modest at best.

Chapter 8 begins with a characterization of limestone-tempered pottery from three Southeastern sites, Icehouse Bottom, Tunacunnhee, and Leake, which met two conditions: (1) the local manufacture of limestone-tempered pottery could be accepted without ambiguity and (2) evidence of Hopewell interaction was present. After the compositions of ceramics from these sites had been determined, the second part of Chapter 8 was devoted to a comparative analysis of limestone-tempered pottery from the Southeast and the Ohio Valley in search of evidence of cultural interaction. The results of this comparison reinforce the view that most of the limestone-tempered vessels recovered from Ohio Hopewell sites were made locally. One notable exception, a limestone-tempered vessel from the Turner site, appears to be a Southeastern import into Ohio.

Chapter 9 revisits the petrographic analysis of ceramic thin sections from Pinson Mounds that had been published previously (Stoltman and Mainfort 2002). The 2002 study was unique at the time, because it involved the compositional analysis of several vessels that had previously been subjected to neutron activation analysis (Mainfort et al. 1997) and came to contradictory conclusions. Whereas the NAA study concluded that "all analyzed pottery from Pinson Mounds was produced locally" (Mainfort et al. 1997:43), the results of the petrographic analysis indicated that "a significant minority of the sampled vessels were imported" (Stoltman and Mainfort 2002:28).

The original petrographic analyses were based solely upon bulk compositions. In Chapter 9 mineralogical compositions are added, providing significant new data to the discussion and allowing a more substantive reconsideration of the petrographic evidence from the site.

In many ways Chapter 10 is the culmination of this study, for it is here that comparative analyses are presented to evaluate the nonlocal status of the 85 vessels from the Ohio Valley sites originally set aside as "Possibly Nonlocal" and to determine their origins if possible. Not included in the original "Possibly Nonlocal" vessels, but added in this chapter for consideration with them, are 10 sherds from 5 northern Kentucky sites that were provided to the author for evaluation as probable intrusives from the Southeast. Following the exclusion of 30 vessels reassessed as local (this includes 23 of the limestone-tempered vessels and 7 unusual vessels), the number of "Possibly Nonlocal" vessels remaining to be evaluated was reduced to 65. This total was further reduced to 40 after sources for 25 of these vessels could not be determined.

Along with the documentation of the occurrence of nonlocal vessels in the centers of the Ohio Valley, an important goal of the analysis of the ceramics from the Southeastern sites was to search for evidence of intrusive vessels at these sites as well. As it turned out, 21 vessels from Southeastern sites were demonstrated to be of nonlocal derivation with 6 of these from undetermined sources. The net total, then, of intrusive vessels with identified sources in this study is 55.

A surprising outcome of the sourcing of pottery vessels identified as foreign in this study was the prominence of the Leake and Mann sites. Of the 55 vessels for which sites-of-origin were determined with a reasonably high degree of probability, 32 (58 percent) were traced to the Leake site and 7 to Mann. Two vessels recovered at Icehouse Bottom and one from Mann were the only vessels identified as probable exports from the Ohio Hopewell centers. Beside their apparent prominence, there is also evidence of direct interaction between the Mann and Leake sites in that sherds from each site were recovered at the other. It appears that these two sites enjoyed special roles in the operation of the Hopewell Interaction Sphere.

Two important observations pertaining to the 55 intrusive vessels from known sources (40 from the Ohio Valley plus 15 from the Southeast) contribute significantly to understanding the nature of the cultural interaction involved. First, there is a pronounced asymmetry in the geographic distribution of sites exporting vessels: 45 were derived from the Southeast, while 10 originated in the Ohio Valley region (7 from Mann and only 3 from Ohio Hopewell sites, one of which went to Mann). This distribution is even more lopsided if the 31 vessels from undetermined sources are considered, for virtually all of these (based upon their mineralogy) originated in the Southeast. Second, with but two exceptions, all of the vessels recovered from sites other than their place of origin were utilitarian, not special purpose, vessels: the preponderance are simple stamped, cordmarked, and plain (i.e., everyday) dishes. From these two observations it is reasonable to infer that these were not "trade" items *per se* within the Hopewell Interaction Sphere. Rather, they are better viewed as travel gear used for food preparation by the travelers on their lengthy journeys to and from the Hopewell centers of the Ohio Valley region. Loss or inadvertent breakage while resident at their destinations would easily account for the occurrence of these vessels at various Hopewell sites.

Another implication of the asymmetric distribution of the foreign vessels is that most of the travelers were Southeasterners who were attracted for a variety of reasons to the great Hopewell centers of the Ohio Valley region. One limitation of this view of the raison d'être for these foreign vessels in the Ohio centers is that their presence sheds little direct light on the purposes or goals of the travelers. One can only speculate on these (e.g., Carr 2005).

The number of Southeastern visitors to the Ohio centers was probably not great and their stays were probably relatively brief because there is no archaeological evidence of whole Southeastern assemblages in the Ohio Valley region suggestive of the presence of groups of people residing at a place that, for example, might be identified as a trade mission. Recurrent visits by small groups of Southeasterners for brief periods could easily account for the existence of the simple-stamped and check-stamped vessels of the Turner series, which could have been made by Southeasterners during their presence in Ohio after their own pottery had been broken or lost. In regard to pottery made from local materials that bears nonlocal decoration or surface treatment, the Mann site stands out as unique. For the site possesses more complicated stamped pottery (literally hundreds of sherds, Kellar 1979:103) than any site beyond the Swift Creek heartland, yet with a single exception, all of the complicated stamped vessels from the site that were thin sectioned were made from local materials (Ruby and Shriner 2005). Considering that the Leake site has a major Swift Creek Component and the direct evidence of two-way contacts between Leake and Mann, the prominence of complicated stamped pottery at Mann can be viewed as further evidence of an intense and perhaps enduring relationship between the two sites.

1
Methodology

Thin Sections and the Petrographic Microscope

Petrography is a venerable geological technique that provides reliable identifications of minerals and rocks. In this role it is well suited for the analysis of pottery because, as a mixture of mineral and rock fragments, most of which are natural inclusions but some of which may also be intentional human additives (i.e., temper), pottery can effectively be treated as if it were a metamorphosed sedimentary rock (Williams 1983:302). Knowledge of these minerals and rocks is valuable for archaeological interpretations concerning the production, use, and exchange of pottery vessels, all topics that will be addressed in the following chapters.

One liability of the technique is that it is destructive, requiring the sacrifice of a slice of a sherd that is then prepared and mounted on a standard glass petrographic slide. The glass slides measure 27 mm x 46 mm, which thus constrains the size that a thin section can be. Normally, sherds are sliced at right angles to the vessel wall, which means most thin sections measure less than 10–15 mm in breadth (depending upon the thickness of the vessel walls) and can be no longer than the maximum length of the slide. Sherds can be cut transversely, i.e., parallel to the vessel walls, thus producing a larger thin section but at the sacrifice of a greater portion of a sherd. For small sherds—say smaller than a quarter-dollar in maximum diameter—little if any of a sherd will survive. But, for larger sherds, a significant portion may be returned to the original donor with only minor damage or retained as a "voucher" for future reference.

To prepare a sherd for mounting on a glass slide, it is first cut to create a flat surface that is further refined by polishing. An epoxy is then used to attach the polished sherd to a glass slide. After the epoxy cures, the sherd is again cut parallel to the initial cut, resulting in a "thick section" (perhaps 1 or 2 mm thick) now permanently attached to the slide. In the final step, one requiring considerable experience and skill, the thick section is ground down to a standard thickness of 30 microns, i.e., .03 mm. The reason for this latter, important step is that the behavior of light passing through a mineral (which is essential for making identifications) is dependent upon two variables, the thickness and the character of the mineral. Thus, in order to facilitate mineral identification, the thickness of the thin section is standardized. All literature describing the optical properties of minerals (e.g., Phillips and Griffen 1981) are premised on 30 microns as the thickness of thin sections. Once mounted, the thin section is usually covered with a protective cover glass and is now a permanent record.

All of this care in the preparation of a thin section is required for its effective analysis on the horizontal, rotatable stage of the petrographic microscope. Unlike biological microscopes, which illuminate specimens with incident light from above, the light source for petrographic microscopes is below the stage from where it is transmitted up through the thin section to the eye of the observer.

In addition, there are several accessories to a petrographic microscope that are essential for its effective use. Especially important are two polarizing lenses (also called nicols after their inventor, William Nicol), one permanently mounted below the stage (the polarizer) and another located above the stage (the analyzer) that can be inserted or withdrawn by the analyst.

In observing the thin sections the analyst works both with the analyzer withdrawn from the field of view (i.e., the light passing through the thin section derives solely from the polarizer and is referred to as plane-polarized light (PPL), because it is vibrating in a single plane dictated by the polarizer) and with the analyzer inserted, i.e., under crossed polars (commonly referred to as crossed nicols and abbreviated XP). When viewing thin sections under plane-polarized light, the natural colors of minerals are visible, which is extremely informative. By contrast, under crossed polars significant optical properties appear (that are complicated to explain) that involve major color changes (so-called interference colors) that are also important for identifying most minerals. As will be seen in later chapters, photomicrographs of thin sections will be shown under both plane and crossed polarized light, for each has special, complementary properties (Figure 1.1).

A second liability of petrography is that it is incapable of identifying individual clay minerals because of their small grain sizes (i.e., <.002 mm). On pottery thin sections, clays appear as a homogenous matrix whose composition can be identified no further. Thus, the focus of petrographic analysis is upon individually visible grains of silt size (>.002 mm) and larger.

A third characteristic (liability?) of petrography is that it deals directly with

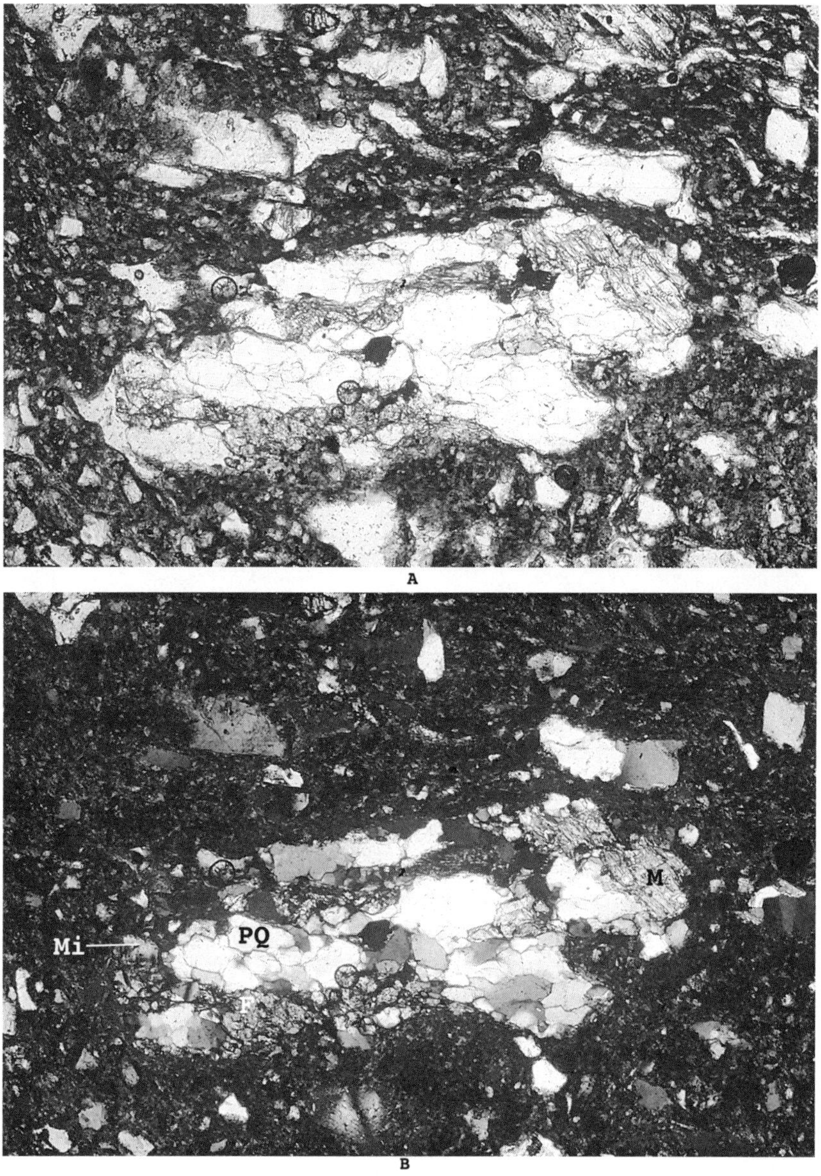

1.1. Two views of thin section 33-286 (at 10X) from the Turner site showing differences between views under plane-polarized light (A) versus crossed polars (B). The rock shown (its maximum dimension=1.675 mm) is a gneiss (or metagranite) composed of polycrystalline quartz [PQ] (uncolored and deformed—note elongation of quartz grains) accompanied by muscovite mica [M] (with high-order interference colors at right end) and microcline [Mi] (feldspar with tartan twinning just showing at the top of the left end under crossed polars). Altered feldspars denoted by F. Feldspar twinning and interference colors are visible only under crossed polars.

the physical properties of rocks and minerals, thus does not provide chemical compositional data. Inferences are sometimes possible as to the chemical makeup of a grain, but this is based solely upon the physical properties observed, not by direct determination, as, for example, is possible with neutron activation or scanning electron microscopy.

The petrographic analysis of pottery came to worldwide attention through the pioneering work of Anna O. Shepard (1936, 1939, 1956). In two studies involving pottery from prehistoric villages in New Mexico—Pecos Pueblo east of Santa Fe (Shepard 1936) and Pueblo Bonito in Chaco Canyon (Shepard 1939, 1954)—Shepard recognized much of the local pottery contained nonlocal igneous rock tempers. She suggested that such pottery was produced at specialized workshops elsewhere, and imported, views that were counter to the prevailing archaeological wisdom of the time that saw such Pueblo communities as economically self-sufficient. These early studies focused entirely on the qualitative identification of tempers, seeing them as exotic based upon the known geological distribution of the rocks involved. One of the arguments used against her views was the possibility that the rocks, not the vessels, were imported—the pots-versus-temper-importation issue that Neil Judd, the excavator of Pueblo Bonito, felt "still hangs in midair" (Judd 1954:235) after an exchange of letters with Shepard.

Always the cautious scholar, Shepard regarded her observation of temper similarities between Chaco Canyon and the Chuska Mountains around 70–80 km to the west as "only circumstantial evidence, not proof" and suggested that a more comprehensive analysis be conducted comparing the pottery from the two areas that involved various stylistic considerations *and* consideration of the type of clays (Shepard 1954:236–237). I mention this situation because it was this expressed need by Shepard for more comprehensive analysis than temper-only observations in petrographic analyses that inspired my effort to introduce a quantitative component to such studies. Later, I was privileged to be able to apply the quantitative approach I was developing to Shepard's very own thin sections from Pueblo Bonito and the Chuska Mountains and, I believe, confirm her importation-from-the-Chuskas hypothesis (Stoltman 1999a).

Point Counting: A Quantitative Approach to Thin Section Analysis

For this study, all thin sections were analyzed using a basic point-counting procedure as described in several earlier publications (Stoltman 1989, 1991, 2001). The goal is to characterize the physical composition of thin sections in both quantitative as well as qualitative terms for use as independent evidence, along with stylistic and contextual considerations, to provide a robust data set for addressing issues of ceramic production and exchange.

The procedure of point counting is analogous to superimposing a rectangular grid over a thin section and recording the minerals that appear at every point of intersection on the grid. The actual procedure, however, is different and less prone to error. It requires a petrographic microscope specially equipped with two properties: (1) a measuring eyepiece with a crosshair and (2) a stage attachment that permits the movement of the thin section back and forth at fixed intervals across the stage beneath the crosshair. With a microscope equipped in this way, the analyst does not have to keep track of where in the grid observations are being made, but only concentrate on identifying whatever appears beneath the crosshair at each stop of the thin section as it is moved across the stage. The main goals of point counting are to provide unbiased estimates of the constituents of a thin section expressed either as a percentage of volume or as a percentage of individual grains.

At the outset it must be emphasized that point counting (referred to as modal analysis in the geological literature, e.g. Chayes 1956; Bayly 1960), is a sampling procedure that is subject to a number of sources of error. Perhaps most important is the issue of sampling error, i.e., the relationship between the sample and the population from which the sample was derived. In dealing with pottery, a thin section is clearly a sample of a vessel, which in turn is a sample of a ceramic assemblage, the latter normally being the target population about which compositional information is being sought. One comforting thought about how reliably a pottery thin section may represent the composition of a vessel is the normal practice of potters to carefully and extensively knead together the ingredients that go into the formation of a vessel. In an empirical test of the proposition of vessel homogeneity, four thin sections from a single vessel were point counted two times each (Stoltman 1989:150–153). The greatest observed compositional difference from the grand mean of the eight counts (presumably the best estimate of the vessel composition) was 2.3 percent (Stoltman 1989:153), which is consistent with the view that thin sections of pottery vessels can usually be accepted as unbiased indicators of vessel composition. Nonetheless, the possibility that a thin section derives from a portion of a vessel that is atypical in some way (for example, has a unique grain or was poorly mixed) must always be kept in mind.

Counting error is another possible source of error in point counting. This source of error is defined as the differences between two or more counts of the same sample (Chayes 1956:59–60; Stoltman 1989:152). Counting error may also be viewed as a measure of precision, which "refers to the size of deviations from the mean m obtained by repeated application of the sampling procedure" (Cochran 1963:16). In an effort to evaluate this error source in the point counting procedure used in this study, 25 pottery thin sections were point counted 2 times each, with 6 of these counted a third time (Stoltman 1989:153–155). The maximum amount of counting error for any pair or triad of observations was less than 5 percent so that the level of precision obtainable with this procedure can be expected to be 95 percent or better (Stoltman 1989:153–155).

Point counting pottery sherds is not precisely analogous to point counting geological samples. It is common for geologists to count several hundred to over one thousand points, but this involves using multiple thin sections derived from rocks that may cover several square kilometers. By contrast, archaeologists are hard pressed to obtain more than a single sherd from a vessel, which means that the thin section area available for counting is severely limited. Accordingly, certain accommodations have to be made.

A thin section may reasonably be regarded as a random sample of a pottery vessel (really, there is no alternative), but the actual points counted cannot be truly random and at the same time occur in large enough numbers over the delimited area of a pottery thin section to have any reasonable expectation of attaining a statistically reliable sample. Thus, rather than attempting a random sample, the point counting procedure used here following Chayes (1956:11) is best described as systematic sampling, which he feels "will usually be more precise than a simple random one" (Chayes 1956:12).

Through trial and error, a compromise-sampling scheme was devised that yields consistent results. A sampling interval of 1 mm is used. It has proven to be the largest practical interval for most pottery thin sections, yet is small enough to ensure a minimum count of 200 points in most cases. In the early years of this study the normal practice was to count until the total area of a thin section had been covered or until 200 points, excluding voids (i.e., cracks or flaws), had been reached, whichever came first. In more recent years a minimum of 200 points was always counted, reversing the slide if necessary and counting in the opposite direction in order to reach the 200 points.

Experience has shown that by employing the 1-mm counting interval, it is possible to compile simultaneously both volumetric and individual-grain data for pottery thin sections. For such variables as relative amounts of sand or silt or temper, etc., volume of such inclusions is the most appropriate measure. It is here that the so-called Delesse relation applies, i.e., that the area of a mineral provides an unbiased estimate of its volume in a thin section (Chayes 1956:4–15). The 1-mm interval allows the analyst to keep track of any multi-count grains that occur. Indeed, this must be done for volumetric estimates to be valid. With a smaller counting interval, pottery thin sections with multiple grains of coarse sand and even gravel sizes (a common property) will become virtually impossible to count reliably. By contrast, individual grain counts may also be recorded simply by keeping track of multi-count grains and counting them as one when, for example, compiling size indices for sand and temper.

The result of this approach to point counting is a physical characterization of each thin section in terms of the percentages of silt-size inclusions plus the species, sizes, and percentages of mineral inclusions of sand size and larger. Since individual clay particles are not identifiable in thin section, their presence is recorded simply as "matrix." The point counting was routinely conducted "blind," that is, without prior knowledge of the samples other than their thin section number. The data for individual thin sections were neither tabulated nor compared until the analysis of each data set had been completed.

Six indices are used to express the results of the point counting: (1) bulk composition; (2) a mineralogical index that accompanies bulk composition; (3) body; (4) paste (these four are volumetric measures); (5) sand-size index; and (6) temper-size index (the latter two are grain-count measures). Bulk composition and its accompanying mineralogical index plus the sand-size indices are used to characterize those vessels that lack visible or reliably identifiable temper, i.e., are untempered or sand tempered (for the latter it is normally impossible to determine objectively which sand grains are natural inclusions and which are human additives).

Bulk composition is a volumetric measure comprised of the relative percentages of three variables, matrix/clay, silt, and sand (always understood to include gravel when present). These are all purely size grades as follows: matrix/clay=<.002 mm; silt= visible grains ranging in size from .002 mm to .0624 mm; and sand/gravel= grains larger than .0625 mm in maximum diameter. As a complement to this index, a sand-size index is also recorded for each vessel. This is devised by assigning a value within the following ordinal scale (based on the Wentworth Scale; e.g., Rice 1987:38) to each sand-sized (or larger) grain encountered in point counting based upon its maximum diameter:

1. Fine=.0625–.249 mm
2. Medium=.25–.499 mm
3. Coarse=.50–.99 mm
4. Very coarse=1.00–1.99 mm
5. Gravel=2.00+ mm

The size values for each of the grains were then summed and divided by the total number of grains counted, thus providing a mean "sand-size index," which ranges between one and five for each thin section.

In addition a mineralogical index will be utilized for vessels that are characterized by bulk composition. This index is comprised of three variables: percent monocrystalline quartz; percent polycrystalline grains; and percent FMM (i.e., feldspars, mafics [i.e., dark minerals such as amphiboles and pyroxenes], and micas). This index is derived from counts on all sand and larger grains whose mineralogy was recorded during point counting. Care is required in using this index because the number of grains counted for such vessels typically range between 50 and 100. Thus, for such small samples the relative percentages among the three classes are prone to sampling error. If used cautiously, this index is useful evidence to enhance our understanding of the mineralogical makeup of thin sections characterized by bulk composition.

By contrast, for vessels with humanly added tempers, the paste/body/temper-size indices will be employed (see Stoltman 1991:109–110; 2001:314). Paste determination involves first identifying temper grains, which are then temporarily set aside until the body index is compiled. The result is represented quantitatively as the percentages of the same three natural ingredients—matrix/clay, silt, and sand—that characterize the bulk composition of raw sediments and untempered vessels. A size index is also recorded for the naturally occurring sand grains in each thin section. Because paste is isomorphic with bulk composition (which applies to both untempered vessels and raw sediments), meaningful, direct comparisons among these three classes are possible. Paste, then, is the appropriate index for assessing the local manufacture of both tempered and untempered ceramic products because it can be compared reliably to local sediments. When thin sections of a representative sample of local sediments are unavailable, "local" paste characteristics can still be defined, albeit tentatively, from a sample of "local" vessels so identified by archaeologists familiar with the local archaeology. Once the paste characteristics of local pottery are determined, comparisons with stylistically anomalous vessels can be made via the so-called Local-Products-Match Postulate (see Stoltman 2001:316–317) to assess more objectively their possible nonlocal origin. The case for the nonlocal derivation of a vessel can be strengthened if its paste not only differs from that of local vessels but also matches that of sediments and/or vessels from a suspected external source.

The identification of humanly added tempers is critical to the definition of body indices. Temper has sometimes been defined solely on the basis of size. For example: Shepard (1936:410) counted only sand-size grains as temper; Maggetti (1982:123) suggested that temper be defined as all grains with a diameter greater than .015 mm (i.e., coarse silt and larger); while Schubert (1986:164) recognized temper as non-plastic mineral grains in excess of .1 mm (i.e., fine sand and larger). In this study size-alone as a definition of temper is rejected because of an abundance of evidence from thin sections of natural sediments that both silts and sands are common inclusions. The identification of some tempers, such as grog, limestone, and coarse grit (i.e., crushed rocks other than limestone) can be relatively straightforward. For vessels with sand temper or finely crushed grit, temper identification is so problematic that when these are encountered, bulk composition rather than the paste/body indices will be used. For sand grains in clearly tempered vessels, it is sometimes difficult to distinguish natural inclusions from smaller temper fragments, especially when dealing with monocrystalline quartz. In these cases, angularity and evidence of igneous or metamorphic alteration (i.e., undulose extinction) are the usual deciding factors, natural sands tending to be more rounded and having straight extinction. It is frankly difficult to associate silt-size particles with most grit tempers, so in such vessels silt grains are usually included as part of the paste (limestone and shell tempers are exceptions in this regard).

Temper type, size, and amount are all variables that directly reflect human actions, thus are likely to be culturally determined. Body indices, then, can provide a valuable basis independent of paste to assess the local vs. non-local status of ceramic vessels, especially when tempers can be traced to known geologic sources as Shepard did so effectively in her New Mexico studies. But temper may be viewed as more than a passive reflection of a traditional way of making pottery. It may also be seen as reflecting active design considerations of the potters, who could use different body recipes for vessels intended to perform different functions (e.g., ritual vs. cooking vs. storage vs. serving, etc.). Thus, body indices offer possible insights into ceramic technology and production.

2

Ohio Hopewell

Pottery recovered from Hopewellian contexts in southern Ohio (Figure 2.1) is traditionally classified using a typology developed by Olaf Prufer in the 1960s (Prufer and McKenzie 1965; Prufer 1968). All sherds selected for thin sectioning in this study were derived from university and museum collections that had been inventoried under this system, so that terminology has been retained throughout this work.

The Prufer system, as we might refer to it for simplicity, partitions Ohio Hopewell pottery into three formal typological series: Scioto, Hopewellian, and Southeastern (Prufer and McKenzie 1965:19–32). Some refinements were later added, including recognition of a provisional fourth series, Miami, that was characterized as "highly tentative" (Prufer 1968:7). Since no sherds assignable to this series were included in the present study, no further references to it shall be made.

Formal types within each of the three series are designated by binomial terms, a place name for the prefix, and a surface finish or decorative descriptor for the suffix. The Scioto series is composed of two undecorated types, McGraw Cordmarked (accepting cordmarking as a surface finish, not decoration) and McGraw Plain, that together dominate the ceramic inventory of all Ohio Hopewell sites (Prufer and McKenzie 1965:19–22). The Hopewellian series consists of several decorated types, all with the Chillicothe prefix, that "are generally qualitatively of a high order. . . . well made, carefully decorated and delicately finished" (Prufer and McKenzie 1965:39). The decoration that characterizes this series consists of various forms, sometimes combinations, of incising, rocker stamping, and brushing (Figure 2.2). Rim sherds with distinctive crosshatched incising are included in the Hopewellian series as a separate category whenever fragmentation prevented associating them with decorated body sherds (Prufer and McKenzie 1965:26–32). An undecorated, polished type, Seip Plain, is also included in the Hopewellian series (Prufer 1968:7–8).

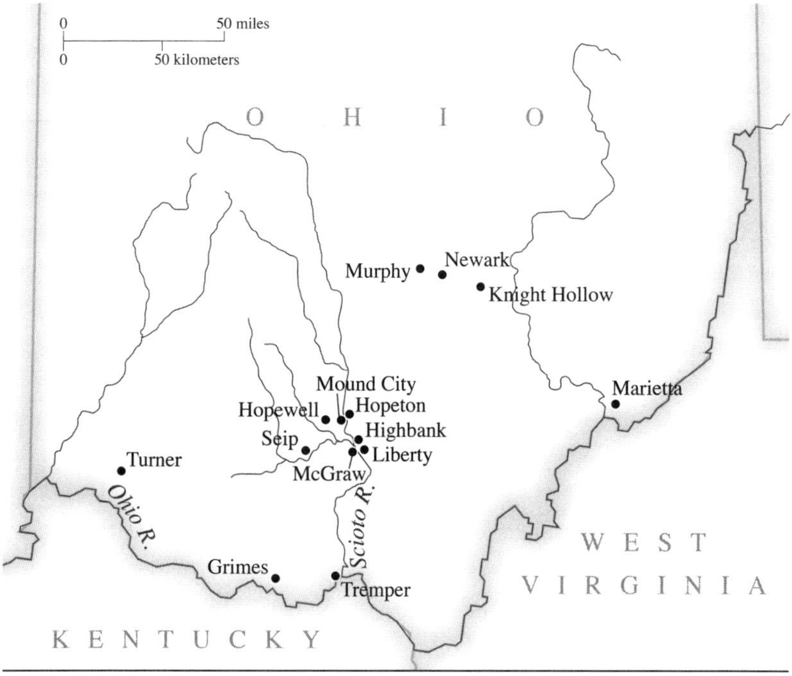

2.1. Map showing locations of all sites in Ohio for which thin section samples were analyzed in this study.

The Southeastern series (with Turner as the type prefix) was defined from a small number of vessels bearing evidence of properties suspected of having been introduced from the Southeastern United States, where they are known to have long histories and rich representations (Figure 2.3); those properties include simple stamping, check stamping, complicated stamping, and vessels supported by tetrapods (Prufer and McKenzie 1965:24–26; Prufer 1968:8–15).

As initially defined (Prufer and McKenzie 1965), the three main Ohio Hopewell ceramic series were essentially undifferentiated on the basis of composition. "Grit" was the predominant or sole temper for all types, but lesser numbers of limestone-tempered vessels were included within both McGraw types in the Scioto series (Prufer and McKenzie 1965:19, 22) as well as some Hopewellian and Southeastern types (Prufer 1968:7, 10–11). Suspicion that some vessels with sand temper were intrusive from the Southeast was expressed in the initial formulation of the Southeastern-series type, Turner Simple-stamped (Prufer and McKenzie 1965:25). This view was later formalized by distinguishing Turner Simple-stamped B from the original grit-

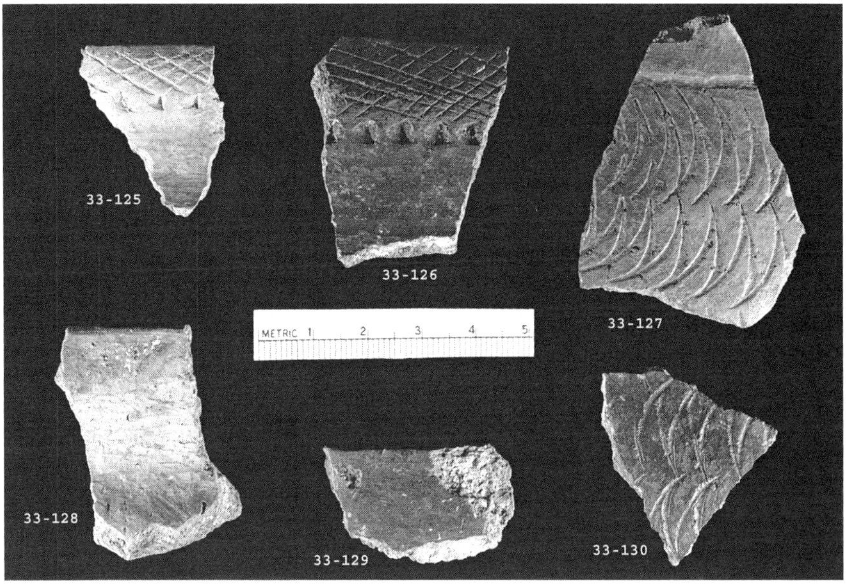

2.2. Examples of Hopewellian-series sherds from the Liberty site (Harness). All were thin sectioned, with appropriate thin section numbers next to each sherd. Nos. 33-125 and 126 are classic Hopewell rims, nos. 33-127 and 130 are Chillicothe Rocker Stamped, no. 33-128 is brushed, and no. 33-129 is burnished.

tempered type, now termed Turner Simple-stamped A (Prufer 1968:8–9), as well as including sand-tempered cordmarked and plain vessels in the Southeastern series with "Untyped" as the prefix (Prufer 1968:14–15).

In defining the three main series, Prufer and McKenzie (1965:18–19) emphasized the local nature of the Scioto series in contradistinction to the nonlocal properties that the other two series shared with Illinois on the one hand (the Hopewellian series) and the Southeast on the other (the Southeastern series). Subsequently, with the exception of sand-tempered vessels, scholars have come generally to treat most types of all three series as products of local Ohio Hopewell potters whether or not they were affected by outside stimuli. In the ensuing thin-section study the issue of local versus nonlocal production will be a point of emphasis through an explicit analysis of vessel composition as a complement to the primarily stylistic observations that characterized the original type definitions.

Thin sections from 225 pottery vessels from 12 Middle Woodland sites plus soil samples from two additional sites in southern Ohio were analyzed for this study (see Figure 2.1). Table 2.1 records the frequencies of thin-sectioned vessels for each of the 12 sites. For each site, the thin-sectioned vessels are further

2.3. Selected Southeastern-series sherds from Seip. All were thin sectioned, with appropriate thin section numbers next to each sherd. Of local origin are the two check-stamped vessels (nos. 33-178 and 185) plus two simple stamped vessels (nos. 33-182 and 183). Apparent Southeast imports are vessels 33-186 and 188 (from Garden Creek), 33-184 (from Icehouse Bottom), and 33-187 (from an unidentified source).

subdivided into the three main typological series—Hopewellian, Scioto, and Turner (i.e., Southeastern)—along with several vessels of possibly nonlocal derivation whose local status will be evaluated later in Chapter 11.

In compiling the data in Table 2.1, a few departures from the standard typology of Ohio pottery were adopted that should be made explicit. These departures were designed to facilitate the application of compositional analysis as an independent variable to the stylistic considerations that informed the standard typology. The essential first step in this analysis of the composition of Ohio Hopewell pottery necessitated the establishment of a reliable compositional baseline for locally manufactured wares. It was only after this step had been taken that the topics of trade and cultural interaction could be effectively addressed pertaining to the possibly intrusive vessels.

In order to maximize the reliability of the sample of vessels employed to determine the compositional signature of local ceramics, only vessels that were both stylistically local *and* grit-tempered were included in the three local series in Table 2.1. This meant that all vessels with limestone and sand

Table 2.1. Frequencies of Thin-Sectioned Sherds by Ceramic Series from Ohio Hopewell Sites

	Hopewellian Series	Scioto Series	Turner Series		Possibly Nonlocal		Totals
			Simple St	Check St	LS	Other	
Chillicothe Region							
Liberty	13	18	2	5	4	5	47
Mound City	6	19	1		2	7	35
Hopewell	3	8		2	—	4	17
Seip	14	16	3	6	3	14	56
McGraw	1	5					6
Eastern Ohio							
Marietta		8			1	1	10
Knight Hollow				1			1
Murphy						3	3
Newark		2				2	4
Southern Ohio							
Tremper	3	4		2			9
Grimes						1	1
Southwestern Ohio							
Turner	4	7	1	1	8	15	36
Totals	44	87	7	17	18	52	225

Note: St=Stamped; LS=Limestone temper

tempers were relegated to the "Possibly Nonlocal" column so that their local status could be independently investigated. Two considerations in particular prompted this decision: (1) these tempers, if present at all, are distinct minorities at all sites in this study except Turner, where limestone tempering was common; and (2) both of these tempers occur widely in the Southeast, raising the distinct possibility of their intrusiveness to Ohio.

A more complicated situation arose with the recognition of grog temper, something not previously reported for Ohio Hopewell pottery (e.g., Griffin 1945; Prufer and McKenzie 1965; but see Prufer 1968:46). Where such vessels were minority types, as at the Liberty and Hopewell sites, their inclusion in the "Possibly Nonlocal" column was straightforward. However, grog temper proved to be predominant at Tremper, and since all nine vessels in that site's sample conformed stylistically to the three main Hopewell-culture series of Ohio, all were considered "local" (see Table 2.1).

As a result of the foregoing considerations, 70 of the 225 vessels were relegated to "Possibly Nonlocal" status in Table 2.1. The assessment of these vessels should be regarded as tentative. After a reliable local compositional baseline has been established for the Ohio Hopewell sites and for several Southeastern sites, the status of these vessels will be reevaluated empirically in Chapters 8 and 10 through comparisons with both local Ohio pottery *and* with potential external sources in the Southeast.

In the following discussion, limestone-tempered vessels that conform to traditional Ohio Hopewell stylistic types will be treated somewhat differently from the remainder of the vessels assigned to the "Possibly Nonlocal" category. Consideration of the 52 "Other" vessels listed in Table 2.1 (i.e., those with sand and grog tempers) will be deferred to Chapter 3 (for grog temper) and Chapter 10 (for vessels with sand or fine grit tempers). By contrast, the paste values of the limestone-tempered vessels can be reliably compared with those of the local, grit-tempered vessels at their respective sites, because problems associated with differential breakage of rock tempers that can bias body values can be set aside. An evaluation of the paste properties of these vessels (excluding body properties) will be conducted at the end of this chapter. Later, in Chapter 8 the limestone-tempered vessels from Ohio will be reevaluated in the context of the limestone-tempered vessels from the Southeast in order to consider possibilities of interaction. In this context, body values will also be used because all the vessels involved (18 from Ohio, 5 from Mann, 1 from Rutherford, and 29 from four Southeastern sites) share the same temper, limestone.

As can be seen from Table 2.1, the number of thin-sectioned vessels from five of the twelve sites is small (i.e., six or fewer), raising legitimate doubts about their representativeness. Moreover, in the cases of four of these sites—Grimes (Brose 1982:31), Knight Hollow (Felumlee 1983), Murphy (Dancey 1991), and Newark (Lepper 1988; Lepper and Yerkes 1997)—their inclusion in the study was specifically to evaluate the status of suspected nonlocal vessels. Thus, in order to attain the initial goal of establishing a reliable compositional baseline for local Hopewell pottery in Ohio, attention will be focused first upon the eight sites with six or more thin-sectioned vessels. Since these sites are scattered across southern Ohio from the Muskingum Valley in the east to the Miami Valley in the west, it was deemed important to analyze them not as a group but in regionally distinct subsets in order to seek evidence of regional variation in ceramic composition within Ohio Hopewell culture.

Discussion will commence with the region around Chillicothe near the confluence of Paint Creek and the Scioto River, long "recognized as the center of the most flamboyant Hopewellian expression anywhere in the Eastern Woodlands" (Ruby, Carr, and Charles 2005:146). Five of the sites in the thin section sample—Liberty, Hopewell, Mound City, Seip, and McGraw—are located in this region providing the richest and most diverse sample for this study. Thereafter, discussion will proceed to eastern Ohio, focusing on the three sites that have locally made ceramics, the Marietta earthworks, the Newark earthworks, and the Knight Hollow Rockshelter; discussion of the Murphy site will be deferred to Chapter 10 because no local pottery was thin-sectioned from the site (see Table 2.1 and Figure 2.1). Next, discussion will proceed to southern Ohio, to the Tremper site near the confluence of the Scioto and Ohio rivers in Scioto County (Grimes will be considered in Chapter 10), and finally to the Turner site in southwestern Ohio (near the confluence of the Little Miami and Ohio rivers in Hamilton County).

Local Pottery of the Chillicothe Region

Five of the sites in this study are located in what is the veritable Hopewell heartland, the region centering on the confluence of Paint Creek with the Scioto River near Chillicothe in Ross County. Four of these sites—Liberty, Hopewell, Mound City, and Seip—are major earthworks that have been the scene of recurrent excavations since the time of Squier and Davis (1848; excellent maps of these sites appear as Plates X, XIX, XX, and XXI in this classic report). By contrast, the fifth site, McGraw, is a small, residential hamlet that was excavated in 1963 (Prufer 1965).

Most of the sherds from the Chillicothe region analyzed in this study were derived from the extant collections of the Ohio State Historical Society, but others were obtained from the Cleveland Museum of Natural History and

Kent State University. A total of 161 sherds from as many vessels were thin sectioned from the five Chillicothe-region sites (Table 2.1). Thirty nine of these vessels (nine with limestone temper) were relegated to the "Possibly Nonlocal" category leaving 122—those that were grit tempered and stylistically local—to form the primary basis for determining the composition of Chillicothe-region ceramics. The compositions for each of these 122 vessels are recorded in Tables 2.2 to 2.12.

To assist the reader in making sense of the data recorded in these tables, some words of explanation are provided here. The tables are organized by *site* and, within each site, by *ceramic series*—Hopewellian, Scioto, and Southeastern (often referred to as Turner for brevity)—in order to facilitate evaluating intra-site and inter-site compositional variation. In each of these tables summary statistical data are also provided, i.e., means and standard deviations for each series for each of the five sites. Additionally, the two-sigma ranges (i.e., 95 percent confidence limits) are recorded for each ceramic series at each site at the bottom of the tables. By comparing individual compositional values for each vessel with the appropriate two-sigma range, vessels with anomalous values (possible intrusives?) are readily identifiable. To ease the reader's burden in this regard, the values that fall outside of the two-sigma ranges are recorded in bold font in the tables and the table cells with the relevant two-sigma values are shaded.

In reviewing the data in Tables 2.2 to 2.12, one cannot help but be impressed with the surprising amount of within-site compositional homogeneity within each of the three ceramic series. The vast majority of compositional values for vessels from the five sites fall within the two-sigma ranges for their respective ceramic series. The relatively few exceptions (denoted in bold font in Tables 2.2 to 2.12) are so minor (usually less than 2 percent) that sampling error within the context of small samples is the most reasonable explanation. The single exception—Vessel 33-91 at Mound City (see Table 2.6)—has unusually high sand percentages in both body and paste suggesting it was made from a distinctive clay. Because no external source can currently be identified for this vessel, it is tentatively accepted as an idiosyncratic local product.

Before turning to the thin section analysis, an additional word of explanation is in order concerning the Liberty site (e.g., Greber 1983; Coughlin and Seeman 1997), what Squier and Davis (1848:56) referred to as "Ancient Work, Liberty Township," because an alternate name—Harness—has also been used for it (e.g., Morgan 1952:89; Prufer 1968:16). Because the site is large and complex, it is confusing to use different names—e.g., Edwin Harness or Russell Brown—for mounds associated with the geometric earth-

Table 2.2. Body and Paste Values by Vessel for the Hopewellian Series at Liberty

Value in bold falls outside the ± two standard deviations/95% confidence limits shown in the shaded cell.

		Body					*Paste*			
Type	**Thin Section No.**	**% Matrix**	**% Sand**	**% Temper**	**Temper Size Index**	**Temper Type**	**% Matrix**	**% Silt**	**% Sand**	**Sand Size Index**
Edwin Harness Mound										
Hopewell Rim	33-125	75	0	25	3.35	Granite	96	4	0	—
Hopewell Rim	33-126	80	1	19	3.14	Granite	95	3	2	1.00
Chillicothe Rocker-St	33-127	77	2	21	3.11	Granite	94	4	2	1.25
Chillicothe Brushed	33-128	73	2	25[22+3]	3.31	Granite+ Gabbro	91	6	3	1.20
Chillicothe Rocker-St	33-130	83	1	16	3.18	Granite	93	6	1	1.50
Seip Plain	33-139	86	2	12	2.50	Granite	94	3	3	1.33
Chillicothe Zoned	33-145	80	3	17	2.41	Granite	92	4	4	2.00
Chillicothe Rocker-St	33-146	78	1	21	2.44	Granite	94	5	1	2.50
Mean ±1 Std Deviation	**[n=8]**	79.0±4.2	1.5±0.9	19.5±4.5	2.93±.41		93.6±1.6	4.4±1.2	2.0±1.3	1.54±.53
Russell Brown Mounds										
Chillicothe Zoned; Md 1	33-243	88	2	10	3.33	Granite	96	1	3	2.00
Hopewell Rim (Md 1)	33-250	81	1	18	3.65	Granite	91	8	1	1.00
Zoned Incised (Md 1)	33-251	87	1	12	2.09	Granite	96	3	1	1.00
Chillicothe Zoned; Md 1	33-252	83	1	16	3.44	Granite	96	3	1	2.00
Chillicothe Zoned; Md 1	33-253	80	0	20	3.78	Granite	**100**	0	0	—
Mean ±1 Std Deviation	**[n=5]**	83.8±3.6	1.0±0.7	15.2±4.2	3.26±.68		95.8±3.2	3.0±3.1	1.2±1.1	1.50±.58
Mean ±1 Std Deviation	**[n=13]**	80.8±4.5	1.3±0.9	17.9±4.7	3.06±.53		94.5±2.5	3.8±2.1	1.7±1.2	1.53±.52
Two-Sigma Ranges	**[n=13]**	71.8-89.8	0-3.1	8.5-27.3	2.00-4.12		89.5-99.5	0-8.0	0-4.1	1.00-2.57

Note: St=Stamped; Md=Mound

Table 2.3. Body and Paste Values by Vessel for the Scioto Series at Liberty

Values in bold fall outside the ± two standard deviations/95% confidence limits shown in the shaded cells.

		Body					Paste			
Type	**Thin Section No.**	**% Matrix**	**% Sand**	**% Temper**	**Temper Size Index**	**Temper Type**	**% Matrix**	**% Silt**	**% Sand**	**Sand Size Index**
Edwin Harness Mound										
McGraw Cordmarked	33-133	82	2	16	3.55	Granite	92	6	2	2.67
McGraw Cordmarked	33-134	76	3	21	2.78	Granite	86	10	4	1.40
McGraw Cordmarked	33-135	78	0	22	3.42	Granite	96	4	0	—
McGraw Cordmarked	33-136	79	2	19	3.39	Granite	92	5	3	1.25
McGraw Cordmarked	33-137	74	1	25	3.56	Granite	91	8	1	1.00
McGraw Cordmarked	33-138	73	2	25 [18+7]	3.03	Granite+ Gabbro	91	6	3	1.75
McGraw Cordmarked	33-147	67	1	32 [27+5]	3.11	Granite+ Gabbro	90	9	1	1.00
McGraw Cordmarked	33-148	73	3	24 [20+4]	3.42	Granite+ Gabbro	89	7	4	1.00
McGraw Cordmarked	33-150	74	1	25	3.77	Granite+	91	8	1	1.00
McGraw Plain	33-129	76	3	21	2.92	Granite	92	5	3	2.75
McGraw Plain	33-140	73	4	23	3.58	Granite+	91	4	5	1.67
Mean ± 1 Std Deviation	[n=11]	75.0±3.9	2.0±1.2	23.0±4.1	3.32±.31		91.0±2.4	6.6±2.0	2.4±1.6	1.55±.67
Russell Brown Mounds										
McGraw Cordmarked; Md 1	33-236	80	3	17	3.58	Granite	95	1	4	1.20

McGraw Cordmarked; Md 2	33-238	73	3	24[14+10]	3.35	Gr+LS-Chert	86	10	4	1.00
McGraw Cordmarked; Md 3	33-241	68	0	32	3.39	Granite	94	6	0	—
McGraw Cordmarked; Md 2	33-248	81	8	**11**[7+4]	3.42	Gr+LS-Chert	84	7	9	1.76
McGraw Plain (Md 3)	33-242	68	8	24	3.42	Granite+	85	5	10	1.82
McGraw Plain (Md 3)	33-245	74	0	26	3.67	Granite	**99**	1	0	—
McGraw Plain (Md 3)	33-247	69	8	23	3.46	Granite+	86	3	**11**	2.06
Mean ±1 Std Deviation	**[n=7]**	73.3±5.5	4.3±3.7	22.4±6.7	3.47±.11		89.9±6.0	4.7±3.3	5.4±4.6	1.57±.44
Mean ±1 Std Deviation	**[n=18]**	74.3±4.5	2.9±2.6	22.8±5.1	3.38±.26		90.6±4.0	5.8±2.7	3.6±3.4	1.56±.59
Two-Sigma Ranges	**[n=18]**	65.3-83.3	0-8.1	12.6-33.0	2.86-3.90		82.6-98.6	0.4-11.2	0-10.4	1.00-2.74

Note: Md=Mound

Table 2.4. Body and Paste Values by Vessel for Turner Check-Stamped and Turner Simple-Stamped A at Liberty

All values fall within the ± two standard deviations/95% confidence limits shown at the bottom of the table.

		Body					*Paste*			
Type	**Thin Section No.**	**% Matrix**	**% Sand**	**% Temper**	**Temper Size Index**	**Temper Type**	**% Matrix**	**% Silt**	**% Sand**	**Sand Size Index**
Edwin Harness Mound										
Turner Check-St	33-132	86	1	13	3.05	Granite	96	3	1	1.50
Turner Check-St	33-141	84	3	13	3.46	Granite	94	3	3	1.50
Turner Simple-St A	33-143	82	1	17	3.21	Granite	92	7	1	2.50
Mean ±1 Std Deviation	**[n=3]**	84.0±2.0	1.7±1.2	14.3±2.3	3.24±.21		94.0±2.0	4.3±2.3	1.7±1.2	1.83±.58
Russell Brown Mounds										
Turner Simple-St A (Md 3)	33-246	78	0	22	2.76	Granite	96	4	0	—
Turner Simple-St A (Md 2)	33-301	78	6	16	3.57	Granite	88	5	7	2.09
Turner Simple-St A (Md 2)	33-302	77	6	17	3.72	Granite	88	4	8	1.67
Turner Simple-St A (Md 2)	33-303	82	0	18	3.24	Granite	95	5	0	—
Mean ±1 Std Deviation	**[n=4]**	78.8±2.2	3.0±3.5	18.2±2.6	3.32±.42		91.8±4.4	4.5±0.6	3.7±4.4	1.88±.30
Mean ±1 Std Deviation	**[n=7]**	81.0±3.4	2.4±2.6	16.6±3.1	3.29±.33		92.7±3.5	4.4±1.4	2.9±3.3	1.85±.44
Two-Sigma Ranges	**[n=7]**	74.2-87.8	0-7.6	10.4-22.8	2.63-3.95		85.7-99.7	1.6-7.2	0-9.5	1.00-2.73

Note: St=Stamped; Md=Mound

Table 2.5. Body and Paste Values by Vessel for the Turner and Hopewellian Series at Mound City

All values for Hopewellian vessels fall within the ± two standard deviations/95% confidence limits shown at the bottom of the table.

		Body					*Paste*			
Type	**Thin Section No.**	**% Matrix**	**% Sand**	**% Temper**	**Temper Size Index**	**Temper Type**	**% Matrix**	**% Silt**	**% Sand**	**Sand Size Index**
Turner Series										
Turner Check-Stamped	33-78	57	0	43	3.40	Granite	94	6	0	—
Hopewellian Series										
Zoned Incised	33-81	82	3	15	2.80	Granite	85	12	3	1.25
Zoned+Bossed	33-82	80	2	18	2.82	Granite	92	6	2	1.67
Zoned Rocker Stamped	33-83/92	84	1	15	2.99	Granite	88	11	1	1.00
Zoned Dentate Stamped	33-84	87	1	12	2.82	Granite	84	15	1	1.00
Zoned Dentate Stamped	33-85	79	2	19	2.94	Granite	92	5	3	1.33
Polished Plain	33-88	80	3	17[15+2]	1.83/2.67	Granite+ LS★	84	12	4	1.75
Mean ±1 Std Deviation	[n=6]	82.0±3.0	2.0±0.9	16.0±2.5	2.76±.30		87.5±3.8	10.2±3.9	2.3±1.2	1.33±.32
Two-sigma ranges	[n=6]	76.0-88.0	0.2-3.8	11.0-21.0	2.16-3.36		79.9-95.1	2.4-16.5	0-4.7	1.00-1.97

★*Limestone identified from angular voids*

Table 2.6. Body and Paste Values by Vessel for the Scioto Series at Mound City

Values in bold fall outside the ± two standard deviations/95% confidence limits shown in the shaded cells.

		Body					*Paste*			
Type	**Thin Section No.**	**% Matrix**	**% Sand**	**% Temper**	**Temper Size Index**	**Temper Type**	**% Matrix**	**% Silt**	**% Sand**	**Sand Size Index**
McGraw Cordmarked	33-64	72	3	25	3.19	Granite	91	5	4	1.00
McGraw Cordmarked	33-65	76	1	23	3.32	Granite	95	4	1	1.00
McGraw Cordmarked	33-67	82	0	18	3.00	Granite	86	14	0	—
McGraw Cordmarked	33-69	77	1	22	3.71	Granite	95	4	1	1.00
McGraw Cordmarked	33-70/71	74	1	25	3.48	Granite	93	6	1	1.00
McGraw Cordmarked	33-96	84	1	15[8+7]	4.17	Gabbro+ Gran	88	11	1	1.00
McGraw Cordmarked	33-68	83	1	16[13+3]	3.77	Granite+ LS*	92	7	1	1.00
McGraw Cordmarked	33-97	88	0	12[10+2]	3.12	Granite+ LS*	95	5	0	—
McGraw Cordmarked	33-98	74	1	25[17+8]	3.14	Granite+ LS*	88	10	2	1.25
McGraw Cordmarked	33-66	71	3	26[24+2]	3.11	Granite+ LS	89	7	4	1.44
McGraw Cordmarked	33-100	78	0	22[12+10]	3.39	Granite+ LS	87	13	0	—
McGraw Plain	33-86	71	1	28	3.10	Granite	93	5	2	1.67

McGraw Plain	33-87	82	1	17	3.06	Granite	92	6	2	1.67
McGraw Plain	33-89	76	1	23	3.70	Granite	88	11	1	1.00
McGraw Plain	33-90	88	3	9	2.92	Granite	88	9	3	1.00
McGraw Plain	33-91	**55**	**12**	33[29+4]	**4.29**	Gabbro+ Gran	**72**	10	**18**	1.42
McGraw Plain	33-93	84	1	15	3.00	Granite	94	4	2	1.00
McGraw Plain	33-94	69	0	31[25+6]	3.60	Granite+ LS*	89	11	0	—
McGraw Plain	33-95	74	1	25[15+10]	3.11	Granite+ LS*	91	8	1	1.00
Mean ±1 Std Deviation	**[n=19]**	76.7±7.8	1.7±2.7	21.6±6.4	3.38±.40		89.8±5.2	7.9±3.2	2.3±4.0	1.16±.26
Two-Sigma Ranges	**[n=19]**	61.1-92.3	0-7.1	8.8-34.4	2.58-4.18		79.4-100	1.5-14.3	0-10.3	1.00-1.68

**Identified from angular voids*

Table 2.7. Body and Paste Values by Vessel for the Hopewellian and Turner Series at the Hopewell Site

All compositional values fall within the relevant ± two standard deviations/95% confidence limits shown below each series.

		Body					*Paste*			
Type	**Thin Section No.**	**% Matrix**	**% Sand**	**% Temper**	**Temper Size Index**	**Temper Type**	**% Matrix**	**% Silt**	**% Sand**	**Sand Size Index**
Hopewellian Series										
Plain, Channeled Rim	33-155	71	6	23	2.94	Granite	87	6	7	2.00
Zoned, Rocker Stamped	33-159	81	3	16	2.14	Granite	86	10	4	1.00
Zoned, Rocker Stamped	33-160	88	0	12	2.05	Granite	96	4	0	—
Mean ±1 Std Deviation	**[n=3]**	80.0±8.5	3.0±3.0	17.0±5.6	2.38±.49		89.6±5.5	6.7±3.1	3.7±3.5	1.50±.71
Two-Sigma Ranges	**[n=3]**	63.097.0	09.0	5.828.2	1.403.36		78.6100	0.612.9	010.7	1.002.92
Turner Series										
Simple-Stamped A	33-157	84	2	14	3.19	Granite	88	10	2	2.50
Simple-Stamped A	33-162	83	0	17	3.69	Syenite	96	4	0	—
Mean ±1 Std Deviation	**[n=2]**	83.5±0.7	1.0±1.4	15.5±2.1	3.44±.35		92.0±5.7	7.0±4.2	1.0±1.4	2.50±0
Two-Sigma Ranges	**[n=2]**	82.1-84.9	0-3.8	11.3-19.7	2.44-4.12		80.6-100	0-15.4	0-3.8	2.50±0

Table 2.8. Body and Paste Values by Vessel for the Scioto Series at the Hopewell Site

Values in bold fall outside the ± two standard deviations/95% confidence limits shown in the shaded cell.

		Body					*Paste*			
Type	**Thin Section No.**	**% Matrix**	**% Sand**	**% Temper**	**Temper Size Index**	**Temper Type**	**% Matrix**	**% Silt**	**% Sand**	**Sand Size Index**
McGraw Cordmarked	33-161	73	1	26	2.95	Granite	89	10	1	2.50
McGraw Cordmarked	33-163	71	4	25	3.47	Granite	85	10	5	2.12
McGraw Cordmarked	33-164	68	2	30	3.13	Granite	89	8	3	1.00
McGraw Cordmarked	33-165	80	4	16	3.05	Granite	88	8	4	**3.00**
McGraw Cordmarked	33-166	90	4	6	3.17	Syenite	87	9	4	**3.67**
McGraw Cordmarked	33-167	65	1	34	3.24	Granite	91	8	1	1.00
McGraw Plain	33-151	82	1	17	3.31	Granite	95	4	1	1.00
McGraw Plain	33-156	77	4	19	3.62	Granite	87	8	5	2.00
Mean ±1 Std Deviation	**[n=8]**	75.8±8.2	2.6±1.5	21.6±8.9	3.24±.22		91.0±5.7	6.0±2.8	3.0±2.8	1.50±.71
Two-Sigma Ranges	**[n=8]**	59.4-92.2	0-5.6	3.8-39.4	2.80-3.68		79.6-100	0.4-11.6	0-8.6	1.00-2.92

Table 2.9. Body and Paste Values by Vessel for the Hopewellian Series at Seip

Values in bold fall outside the ± two standard deviations/95% confidence limits shown in the shaded cells.

		Body					*Paste*			
Type	Thin Section No.	% Matrix	% Sand	% Temper	Temper Size Index	Temper Type	% Matrix	% Silt	% Sand	Sand Size Index
Zoned Rocker Stamped	33-168	82	1	17	3.18	Granite	81	18	1	1.00
Zoned Rocker Stamped	33-169	80	1	19	3.36	Granite	**77**	**21**	2	1.00
Zoned Rocker Stamped	33-170	88	1	11	3.24	Granite	82	17	1	1.00
Zoned Rocker Stamped	33-171	80	0	20	3.53	Granite	83	17	0	—
Zoned Rocker Stamped	33-172	82	1	17	3.58	Granite	86	13	1	1.00
Plain Rocker Stamped	33-174	77	2	21	3.48	Granite	87	10	3	1.00
Plain Rocker Stamped	33-175	88	2	10	3.40	Granite	87	10	3	1.00
Plain Rocker Stamped	33-176	83	2	15	3.38	Granite	85	13	2	1.00
Plain Rocker Stamped	33-177	77	0	23	3.26	Granite	89	11	0	—
Plain Rocker Stamped	33-210	83	0	17	2.94	Granite	85	15	0	—
Dentate Rocker Stamped	33-180	87	1	12[11+1]	3.24/5.00	Granite+ Grog	84	15	1	1.00
Dentate Rocker Stamped	33-181	87	**3**	10	2.75	Granite	86	11	3	1.00
Hopewell Rim	33-112	85	1	14	3.80	Granite	85	14	1	1.00
Hopewell Rim	33-113	78	0	22	3.36	Granite	86	14	0	1.00
Mean ±1 Std Deviation	[n=14]	82.6±3.9	1.1±0.9	16.3±4.4	3.28±.29		84.5±3.0	14.2±3.2	1.3±1.1	1.00±0
Two-Sigma Ranges	[n=14]	74.8-90.4	0-2.9	7.5-25.1	2.70-3.86		78.5-90.5	7.8-20.6	0-3.5	1.00±0

Table 2.10. Body and Paste Values by Vessel for the Scioto Series at Seip

Values in bold fall outside the ± two standard deviations/95% confidence limits shown in the shaded cells.

		Body					Paste			
Type	Thin Section No.	% Matrix	% Sand	% Temper	Temper Size Index	Temper Type	% Matrix	% Silt	% Sand	Sand Size Index
McGraw Plain	33-116	72	2	26	3.67	Granite	83	**14**	3	1.00
McGraw Plain	33-117	68	**7**	25	3.29	Gabbro	85	6	**9**	1.50
McGraw Plain	33-118	70	5	25	3.24	Gabbro	82	12	6	1.63
McGraw Plain	33-119	73	1	26	3.37	Granite	94	5	1	1.00
McGraw Plain	33-120	84	0	16	3.58	Metamorphic	95	4	1	**3.00**
McGraw Plain	33-121	83	0	17	3.37	Granite	93	6	1	1.00
McGraw Plain	33-122	74	2	24	3.60	Granite	91	6	3	1.67
McGraw Plain	33-201	75	3	22	3.27	Granite	87	10	3	2.00
McGraw Cordmarked	33-203	84	1	15	3.46	Granite	95	4	1	1.00
McGraw Cordmarked	33-204	80	2	18[17+1]	3.86	Granite+LS	94	3	3	1.00
McGraw Cordmarked	33-205	72	1	27	3.40	Granite	96	3	1	1.50
McGraw Cordmarked	33-206	80	0	20	**4.33**	Granite	90	10	0	—
McGraw Cordmarked	33-207	82	1	17	3.71	Granite	97	2	1	1.00
McGraw Cordmarked	33-208	71	0	29	3.72	Granite	96	4	0	—
McGraw Cordmarked	33-209	78	1	21	3.83	Granite	93	5	2	1.00
Plain Tetrapod	33-197	72	0	28	3.26	Granite	94	6	0	—
Mean ±1 Std Deviation	**[n=16]**	76.1±5.4	1.6±2.0	22.3±4.6	3.56±.29		91.6±4.8	6.2±3.4	2.2±2.4	1.40±.59
Two-Sigma Ranges	**[n=16]**	65.3-86.9	0-5.6	13.1-31.5	2.98-4.14		82.0-100	0-13.0	0-7.0	1.00-2.58

Table 2.11. Body and Paste Values by Vessel for Turner Check-Stamped and Turner Simple-Stamped A at Seip

Values in bold fall outside the relevant ± two standard deviations/95% confidence limits shown in the shaded cells.

		Body					*Paste*			
Type	**Thin Section No.**	**% Matrix**	**% Sand**	**% Temper**	**Temper Size Index**	**Temper Type**	**% Matrix**	**% Silt**	**% Sand**	**Sand Size Index**
Turner Check-Stamped	33-178	92	1	7	3.93	Granite	88	11	1	1.00
Turner Check-Stamped	33-179	83	1	16	3.77	Granite	85	**14**	1	1.00
Turner Check-Stamped	33-185	69	2	29	2.79	Granite	92	6	2	1.25
Mean ±1 Std Deviation	[n=3]	81.4±11.6	1.3±.5	17.3±11.1	3.50±.62		88.3±3.5	10.3±4.0	1.3±0.6	1.08±.14
Turner Simple-Stamped A	33-182	69	1	30	3.53	Granite	92	7	1	1.00
Turner Simple-Stamped A	33-183	72	2	26	3.40	Granite	91	6	**3**	1.00
Turner Simple-Stamped A	33-191	77	1	22	3.40	Granite+ Gabbro	94	5	1	1.00
Turner Simple-Stamped A	33-194	88	1	11	2.93	Granite	95	4	1	**2.00**
Turner Simple-Stamped A	33-196	74	2	24	3.19	Granite	95	3	2	1.75
Turner Simple-Stamped A	33-202	75	1	24	3.26	Granite	94	5	1	1.00
Mean ±1 Std Deviation	[n=6]	75.8±6.6	1.4±.5	22.8±6.4	3.29±.21		93.5±1.6	5.0±1.4	1.5±0.8	1.29±.46
Mean ±1 Std Deviation	[n=9]	77.7±8.2	1.3±.5	21.0±8.0	3.36±.37		91.8±3.4	6.8±3.5	1.4±0.7	1.22±.38
Two-Sigma Ranges	[n=9]	61.3-94.1	0.3-2.3	5.0-37.0	2.62-4.10		85.0-98.6	0-13.8	0-2.8	1.00-1.98

Table 2.12. Body and Paste Values by Vessel for Hopewellian and Scioto Series at McGraw

All compositional values for Scioto-series vessels fall within the relevant ± two-sigma ranges/95% confidence limits shown at the bottom of the table.

		Body					*Paste*			
Type	**Thin Section No.**	**% Matrix**	**% Sand**	**% Temper**	**Temper Size Index**	**Temper Type**	**% Matrix**	**% Silt**	**% Sand**	**Sand Size Index**
Hopewellian Series										
Chillicothe Rocker-St	33-106	84	1	15[12+3]	2.80/4.25	Granite+LS	87	12	1	1.00
Scioto Series										
McGraw Cordmarked	33-101	93	2	5	2.67	Granite	78	20	2	1.00
McGraw Cordmarked	33-102	94	1	5	3.00	Granite	82	17	1	2.00
McGraw Cordmarked	33-103	85	0	15	3.83	Granite	86	14	0	—
McGraw Plain	33-104	87	5	8	2.86	Granite	78	17	5	2.88
McGraw Plain	33-105	66	0	34	3.42	Granite	95	5	0	—
Mean ± 1 Std Deviation	**[n=5]**	85.0±11.3	1.6±2.1	13.4±12.2	3.16±.47		83.8±7.1	14.6±5.8	1.6±2.1	1.96±.94
Two-Sigma Ranges	**[n=5]**	62.4-100	0-5.8	0-37.8	2.22-4.10		69.6-98.0	3.0-26.2	0-5.8	1.00-3.84

Note: St=Stamped

works that are the site's hallmark (Seeman and Soday 1980). The Edwin Harness mound, the largest of site's 14 mounds, is located inside the earthwork enclosure and has been the scene of recurrent excavations, most recently by the Cleveland Museum of Natural History (Greber 1983; see also Coughlin and Seeman 1997:241). A total of 22 sherds recovered from the Edwin Harness mound, which were taken both from the Ohio State Historical Society and the Cleveland Museum of Natural History collections, are included in the present study (see Tables 2.2, 2.3, and 2.4). In addition 16 sherds recovered from the Russell Brown mounds, which are "located just outside of the Harness Earthworks" (Prufer 1968:139; see maps in Seeman and Soday 1980:99 and Coughlin and Seeman 1997:241) are also included in this study under the Liberty site in Tables 2.2 to 2.4. In doing so I am following Prufer's suggestion that, "Properly speaking, they should be considered part of the Harness site" (Prufer 1968:139). The sherds from the Edwin Harness and Russell Brown mounds are recorded in separate subsets in Tables 2.2 to 2.4 so that those who prefer to consider them as separate sites may view the data from that perspective.

Based upon the analysis of the 122 "local" thin sections, four generalizations may be offered concerning the pottery-making practices of Hopewellian peoples of the Chillicothe region: (1) the temper of choice was crystalline, plutonic, igneous rocks, mostly of granitic composition; (2) in preparing and adding temper, different sizes and proportions were used for Scioto-series (presumably utilitarian) vessels in contrast to Hopewellian-series vessels that were presumably used for more limited or specialized purposes; (3) the clays used to fabricate the pots were derived mainly from local alluvium with nearby lacustrine deposits also possibly utilized; (4) most vessels of the two Southeastern-series types, Turner Simple-stamped A and Turner Check-stamped, are compositionally indistinguishable from vessels of the other two series, thus can be accepted with reasonable likelihood as locally produced. Each of these observations is discussed more fully in turn below.

Tempers Used by Hopewell Potters

The first general property of Chillicothe-region pottery is the occurrence of coarse grains of igneous rocks as the grit temper of choice. Of the 161 vessels from this region that were thin sectioned for this study, 122 (i.e., all those recorded in the Hopewellian, Scioto, and Turner series columns of Table 2.1) are characterized by such temper. Since "bedrocks exposed at the surface in Ohio are all of sedimentary types" (Lamborn 1938:7), there are only two possible local sources for such rocks, either glacial outwash or glacial till. The southern limits of glaciation in Ohio bisect Ross County just south of the

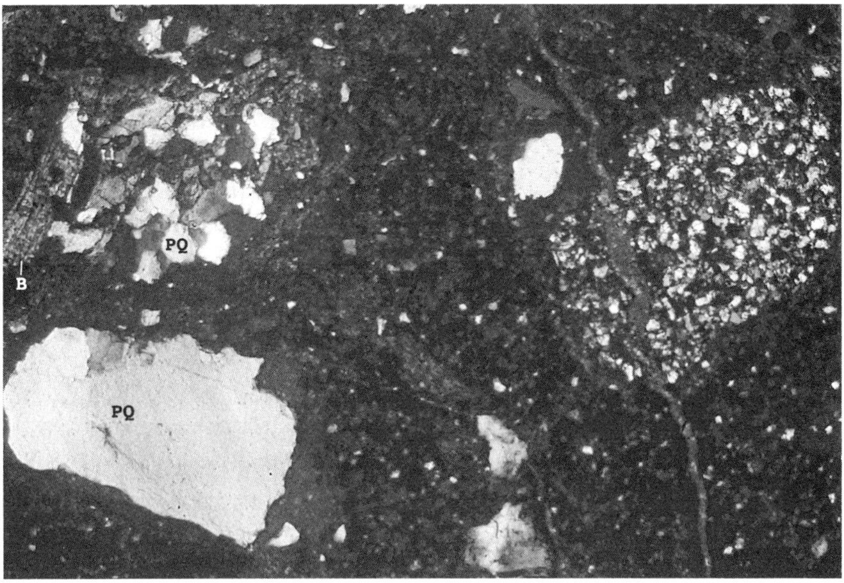

2.4. Photomicrograph of granite temper in Seip vessel 33-201 taken at 10X magnification under crossed polars. B denotes biotite mica. PQ denotes polycrystalline quartz in polymineralic, granite grain and as a separate grain on lower left. Note natural shale grain at upper right= .95 mm.

Paint Creek-Scioto confluence at Chillicothe (Nobel and Korsok 1975:32). With glacial till covering the northern half of Ross County and the Scioto River having served as a major discharge channel for outwash from the nearby glaciers, the potters of the Chillicothe region resided in close proximity to geological deposits that contained a variety of rocks that had been transported from multiple northern locales.

A salient property of glacial till and outwash is the presence of a wide diversity of displaced, transported, and redeposited rock fragments. Since rocks of multiple types occur widely scattered within these glacial deposits, it is unlikely that a discrete temper source analogous to a bedrock quarry can be identified today because any exposures of those sediments accessible to Hopewellian potters have probably long since been buried or destroyed by natural erosion or modern construction activities.

Rocks of granitic composition were, far and away, the prevalent temper employed by the Hopewell potters of the Chillicothe region (Figure 2.4). Defined broadly, granitic rocks are medium to coarse-grained igneous rocks that are composed of 20–60% quartz, with most of the remainder being feldspars (Le Maitre 2002). Lesser amounts of mafic (ferromagnesian) minerals such as micas and amphiboles also occur. Why such rocks were preferentially used is unknown. Perhaps the reason is simply local abundance or accessibility. This is an empirical issue that is in need of field checking. Another possibility is that it was the friability of such rocks that was attractive to the Hopewell potters. Because of the relative abundance of feldspars in granitic rocks, and the propensity of these minerals to undergo chemical decomposition, most granites can be easily crushed into finer particle sizes (thus are often characterized as "rotten"). It is this property that presumably made granitic rocks attractive not only to Hopewellian potters but also to potters widely throughout the world.

Other igneous rocks, including gabbro, diorite, and syenite, along with some sedimentary rocks (e.g., shale, limestone, and sandstone), also were observed as inclusions, usually in minor amounts, in several thin sections. Since the sedimentary rocks occur in bedrock outcrops in southern Ohio, their presence in minor amounts was regarded as natural rather than temper (see Figure 2.4). In several instances, however, limestone (usually as large polygonal voids interpreted as leached limestone) was recognized as the predominant mineral inclusion in a thin section, and in such cases it was treated as temper (Figure 2.5). Because limestone tempering is a common property of pottery in Middle Woodland contexts in the Midsouth (e.g., Chapman 1973:61–68; Jefferies 1976:31–32; Walthall 1980:127), those vessels that were truly limestone tempered (as opposed to possessing an isolated grain or two) were included in the Possibly Nonlocal category for further evaluation.

Differential Temper Sizes and Amounts

The second general property of the Hopewell pottery of the Chillicothe region concerns the differential amounts and sizes of temper in the Scioto and Hopewellian series. Data pertaining to this topic are provided in Table 2.13, which records the mean body values for vessels of each of the three pottery series for each of the five Chillicothe-region sites as well as for all vessels combined from the region for each of the three series.

In evaluating the data on temper percent and size in Table 2.13, it should be stressed that the two variables are not completely independent. Because the body index is a volumetric measure, it can normally be expected that the larger the mean temper size, the greater will be the percent of temper. Nonetheless, a clear trend is evident in Table 2.13; namely, for each of the four earthwork sites, the percent and size of temper in Scioto-series vessels is larger than in the Hopewellian-series vessels from the same site.

The McGraw site, with but a single Hopewell vessel within its small sample of six thin sections, cannot be evaluated objectively with respect to this issue. If the minor amounts of coarse limestone inclusions in this vessel are dis-

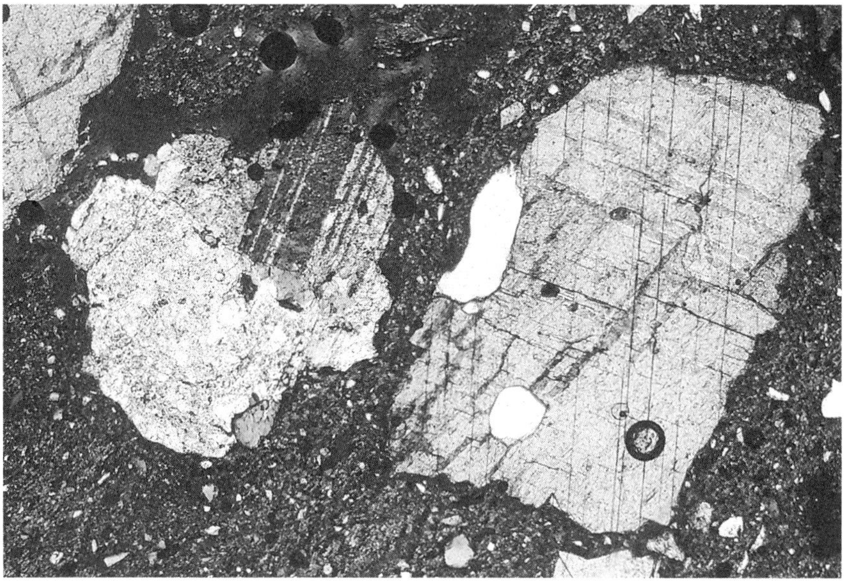

2.5. Photomicrograph of granite and limestone tempers in Vessel 33-100 (McGraw Plain) from Mound City taken at 10X magnification under crossed polars. Limestone grains shown on left and right; granite grain (with tinned plagioclase feldspar) in center=1.05 mm.

counted, the granitic inclusions are smaller in average size (2.80) and volume (12 percent) than comparable values in the Scioto-series at the site (3.16 size index and 13.4 percent volume), which is consistent with the general pattern observed in the other sites.

Despite the apparent within-site compositional differences between ceramic series, the data in Table 2.13 document general between-site similarities among vessels of the same series. For this reason, the body values for vessels of the same ceramic series (i.e., Hopewellian, Scioto, and Turner) were combined to provide a regional estimate of body compositions by series. These values are recorded at the bottom of Table 2.13.

The application of t-tests for significance of differences between the mean values of two of the body parameters, percent Temper and Temper Size Indices, for the Hopewellian versus the Scioto series reveals the following interesting results: t=3.81 for percent Temper yielding a probability of <.001 and t=4.261 for Temper Size Indices yielding a probability of <.001 (both with 101 degrees of freedom). That is to say, the differences between the mean body values for these two series are highly significant at a probability level that approaches infinity. It is fair to infer that the observed compositional differences in body values between the Hopewellian and Scioto series vessels

Table 2.13. Mean Body Values ± One Standard Deviation for Grit-Tempered Hopewell, Scioto, and Turner (Southeastern) Series Vessels for Each of the Five Sites in the Chillicothe Region Plus for All Vessels of the Same Series for All Five Sites Combined

Cells showing temper percentages and sizes for the Hopewellian vs. Scioto series are shaded for easy identification.

Site	Series	N	% Matrix	% Sand	% Temper	Temper Size Index
Liberty	Hopewellian	13	80.8±4.5	1.3±0.9	17.9±4.7	3.06±.53
Liberty	Scioto	18	74.3±4.5	2.9±2.6	22.8±5.1	3.38±.26
Liberty	Turner Ck and Simple-St A	7	81.0±3.4	2.4±2.6	16.6±3.1	3.29±.33
Mound City	Hopewellian	6	82.0±3.0	2.0±0.9	16.0±2.5	2.76±.30
Mound City	Scioto	19	76.5±7.9	1.4±2.4	22.1±6.7	3.38±.38
Mound City	Turner Ck	1	57	0	43	3.40
Hopewell	Hopewellian	3	80.0±8.5	3.0±3.0	17.0±5.6	2.38±.49
Hopewell	Scioto	8	75.8±8.2	2.6±1.5	21.6±8.9	3.24±.22
Hopewell	Turner Simple-St A	2	83.5±0.7	1.0±1.4	15.5±2.1	3.44±.35
Seip	Hopewellian	14	82.6±3.9	1.1±0.9	16.3±4.4	3.28±.29
Seip	Scioto	16	76.1±5.4	1.6±2.0	22.3±4.6	3.56±.29
Seip	Turner Ck and Simple-St A	9	77.7±8.2	1.3±0.5	21.0±8.0	3.28±.21
McGraw	Hopewellian	1	84	1	12 Gran/3LS	2.80/4.25
McGraw	Scioto	5	85.0±11.3	1.6±2.1	13.4±12.2	3.16±.47
McGraw	Turner Ck and Simple-St A	0	—	—	—	—
All Sites	Hopewellian	37	81.7±4.3	1.5±1.2	16.8±4.2	3.05±.47
All Sites	Scioto	66	76.4±7.1	2.1±2.4	21.5±6.8	3.39±.34
All Sites	Turner Ck and Simple-St A	19	78.4±8.1	1.6±1.7	20.0±8.3	3.34±.32
All Sites	Grand Mean	122	78.4±6.9	1.8±2.0	19.8±6.7	3.28±.42

Note: Ck=Check-Stamped; St=Stamped

within the Chillicothe region are attributable to intentional recipe manipulation on the part of Hopewell potters. Because the Scioto series is composed of two undecorated types, McGraw Cordmarked and McGraw Plain, while the Hopewellian series is composed only of decorated types (Prufer 1965:19–32), it is reasonable to view the two series as functionally distinct. The Scioto series presumably was used primarily for such utilitarian functions as storage, transport, and cooking. By contrast, the Hopewellian series likely was used for more limited or specialized functions such as serving, ritual, and display. Thus, the greater volume of temper in Scioto-series vessels presumably enhanced their durability by minimizing the mechanical and thermal stresses associated with their rougher handling and/or recurrent exposure to fire (e.g., Steponaitis 1984). Alternatively, and not necessarily mutually exclusively, the finer temper in Hopewellian-series vessels may be a reflection of the greater care that went into their manufacture that facilitated their efficacious presentation and use for special occasions such as feasts and rituals.

With respect to the differential amounts and sizes of temper that were noted for the Scioto and Hopewellian series, no clear patterns are visible for the two Turner types. As can be seen from the series means recorded at the bottom of Table 2.13, the body values for the Turner series are more variable from site to site than is the case for either of the other two series, although their mean values are generally intermediate between those of the other two series. From these data, it is difficult to discern any evidence of a closer relationship between the Turner vessels and either of the other two series. Perhaps this can be attributed to the apparent cultural hybridization that characterizes this series, or perhaps it is merely an artifact of the smallness of the sample.

Clays Used for Pottery Manufacture

As discussed in Chapter 1, it is possible to estimate the composition of clays used in pottery manufacture by deleting humanly added temper from bulk composition to produce paste indices. The paste values for the 122 "local" vessels provide the most reliable basis available for estimating the compositional character of the clays used by the Chillicothe-region potters (Table 2.14). The mean paste values ± one standard deviation are recorded by ceramic series in Table 2.14 for each of the five Chillicothe-region sites (The values for individual vessels from which these means were compiled are recorded in Tables 2.2 to 2.12). In addition the grand paste means for all 122 local vessels are recorded at the bottom of Table 2.14. Figure 2.6 provides a visual summary perspective on these data that pertains as well to the discussion of the Turner series in the following section.

From Table 2.14 and Figure 2.6, it can be seen that the clays used in mak-

Table 2.14. Means ± One Standard Deviation for Paste Values for Local Vessels by Series from Each of the Five Chillicothe-Region Sites

Shaded cells denote the especially high silt values for Hopewellian-Series vessels from Mound City and Seip.

Site	Series	N	% Matrix	% Silt	% Sand	Sand Size Index
Liberty	Hopewellian	13	94.5±2.5	3.8±2.1	1.7±1.2	1.53±.52
Liberty	Scioto	18	90.6±4.0	5.8±2.7	3.6±3.4	1.56±.59
Liberty	Turner Ck and S-S A	7	92.7±3.5	4.4±1.4	2.9±3.3	1.85±.44
Liberty	All local vessels	38	92.3±3.8	4.9±2.4	2.8±2.9	1.59±.54
Mound City	Hopewellian	6	87.5±3.8	10.2±3.9	2.3±1.2	1.33±.32
Mound City	Scioto	19	89.9±5.2	7.9±3.2	2.3±4.0	1.16±.26
Mound City	Turner Ck	1	94	6	0	—
Mound City	All local vessels	26	89.4±4.9	8.4±3.4	2.2±3.5	1.21±.28
Hopewell	Hopewellian	3	89.6±5.5	6.7±3.1	3.7±3.5	1.50±.71
Hopewell	Scioto	8	91.0±5.7	6.0±2.8	3.2.8	1.50±.71
Hopewell	Turner S-S A	2	92.0±5.7	7.0±4.2	1.0±1.4	2.50±0
Hopewell	All local vessels	13	89.5±3.8	7.6±2.4	2.9±2.2	1.98±.91
Seip	Hopewellian	14	84.5±3.0	14.2±3.2	1.3±1.1	1.00±0
Seip	Scioto	16	91.6±4.8	6.2±3.4	2.2±2.4	1.40±.59
Seip	Turner Ck and S-S A	9	91.8±3.4	6.8±3.5	1.4±0.7	1.22±.38
Seip	All local vessels	39	89.1±5.2	9.2±5.0	1.7±1.7	1.22±.45
McGraw	Hopewellian	1	87	12	1	1.00
McGraw	Scioto	5	83.8±7.1	14.6±5.8	1.6±2.1	1.96±.94
McGraw	All local vessels	6	84.3±6.5	14.2±5.3	1.5±1.9	1.72±.91
All Five Sites	Mean ± 1 Std Dev	122	90.0±5.0	7.7±4.4	2.3±2.6	1.44±.59

Note: Ck=Check-Stamped; S-S=Simple-Stamped; Std Dev=Standard Deviation

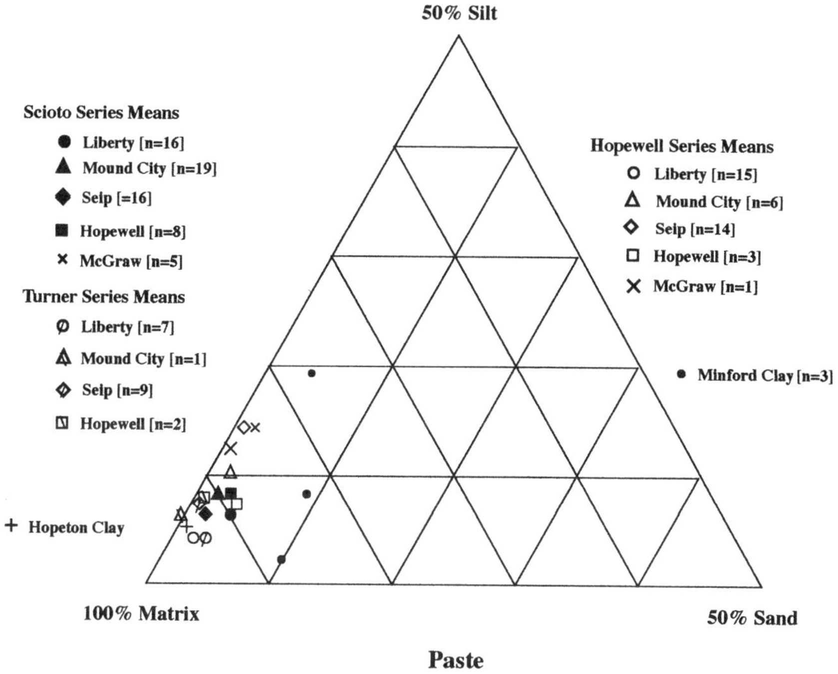

2.6. Ternary graph showing the mean paste values for all Ohio sites by ceramic series plotted against the clay recovered beneath the Hopeton earthworks and the three Minford clay samples.

ing pottery at the five Chillicothe-region sites have generally low amounts of sand (most have less than 4 percent, with a grand mean of only 2.3 percent) and low to moderate, but more variable, amounts of silt (less than 10 percent in most cases). In addition, the sand size indices generally average less than 1.75, even allowing for the occasional occurrence of coarse natural inclusions of shale or limestone. This means that the vast majority of natural, sand-size inclusions are of the medium size grade or smaller (i.e., few are larger than .50 mm).

The relative amount of silt is the most variable paste parameter among the local vessels of the Chillicothe region. At first glance, the occurrence of silty pastes (i.e., those with 10 percent or more silt) appears to be randomly distributed among all sites and within all ceramic series (see Tables 2.2 to 2.12). However, upon closer scrutiny, an apparently nonrandom pattern emerges; namely, the Hopewellian series at Mound City and Seip have relatively high incidences of silt in contrast to the vessels of the other two series at those sites (see the shaded cells in Table 2.14 and Figure 2.6). This pattern is most dramatically evident at Seip where all 14 Hopewellian vessels have 11 percent or more silt (see Table 2.9). It is interesting to consider that the occurrence of these high-silt pastes in Hopewellian vessels reflects intentional choices of the potters, possibly to compensate for the reduction in thermal shock resistance associated with lesser amounts and finer sizes of tempers that characterize this series. If such an intentional practice of ceramic recipe manipulation occurred at Mound City and Seip, one is left to wonder why the potters at Liberty and Hopewell did not share in it.

The McGraw site also deserves mention in this context as five of the six vessels from the site have 12 percent or more silt (see Table 2.12). Since most of these vessels derive from the Scioto series, this appears to be a departure from the "pattern" just discussed. With relatively high-silt pastes present throughout the regional sample, there seems good reason to believe that they were part of the natural variation in local clays that Hopewell potters were free to choose from. To help resolve this issue, comparisons of the paste values with local sediments will be considered.

To gain direct evidence for the compositions of local clays that might have been employed for pottery manufacture in the Ohio Hopewell heartland, sediment samples were collected in and around the following five sites: Seip, Hopewell, Mound City, Liberty, and Hopeton. In addition samples were collected from exposures of a fine-grained, slack-water sediment (derived from glacial Lake Tight) that covers nearly 7,000 square miles in southeastern Ohio (reaching into southeast Ross County) and adjacent parts of Kentucky and West Virginia (Hansen 1987:5). Referred to variously as the Minford clay (Hansen 1987) or silt (Stout and Schlaaf 1931; Ver Steeg 1946), samples of this sediment were collected in Pike, Jackson, and Scioto counties, the three counties directly south of Ross County, because in historic times these clays were known to "have some value for the manufacture of ceramic products" (Stout 1916:430). The bulk composition=paste values for these local sediments, 17 in all, are recorded in Table 2.15 and shown in Figure 2.7.

The sediment samples were collected from four contexts: 1) from half-meter to one-meter deep pits excavated into the B2t horizon (i.e., clay-enriched subsoil) of the alluvial terraces on which the five sites are situated (n=9); 2) in natural exposures in stream banks adjacent to the sites (n=3); 3) in highway or railroad cuts exposing Minford clays (n=3); and 4) from recent excavations beneath two earthworks, Hopeton (n=1) and High Bank (n=1) where what are believed to be slack-water deposits were encountered in the alluvial terraces underlying those sites. Test tiles were prepared for each of these samples, which were then dried, fired, and mounted on glass slides precisely as if they were sherds. Since all of these samples were derived from undisturbed,

Table 2.15. Bulk Compositions/Paste Values for Local Sediments in the Chillicothe Region

Key for sand values: ^{S} *= presence of shale;* ^{L} *= presence of limestone*

Site	Thin Section No.	% Matrix	% Silt	% Sand	Sand Size Index
Seip, Site Terrace B2t	33-226	61	9	30^{S}	2.38
Seip, Creek Bank	33-227	59	8	33^{L}	1.04
Hopewell, Site Terrace B2t	33-223	79	3	18^{S}	1.89
Hopewell, Low Terrace B2t	33-225	52	13	35^{SL}	1.35
Hopewell, Creek Bank	33-224	76	12	12^{L}	1.20
Mound City, Site Terrace B2t	33-220	57	8	35^{S}	3.00
Mound City, Site Terrace B2t	33-221	64	2	34^{S}	2.90
Mound City, Creek Bank	33-222	76	6	18^{SL}	1.43
Liberty, High Terrace B2t	33-230	57	5	38^{S}	2.90
Liberty, Low Terrace B2t	33-231	84	5	11^{SL}	2.26
Hopeton, Middle Terrace B2t	33-229	51	5	44^{L}	1.52
Hopeton, Low Terrace B2t	33-228	78	12	10^{L}	1.27
Hopeton, NPS Below Walls	33-307	94	5	1	1.50
Highbank,CMNH Tr. VI	33-311	88	2	10^{S}	2.48
Minford Clay, Pike Co.	33-304	83	8	9	1.00
Minford Clay, Jackson Co.	33-305	72	9	19	1.17
Minford Clay, Scioto Co.	33-306	77	19	4	1.75

natural sediments (i.e., they lack temper), their bulk compositions can be accepted as equivalent to the paste values of pottery sherds.

By referring to Figure 2.7, it can be seen that 15 of the 17 sediments are too sandy to have been used by Hopewell potters unless the clays were either levigated or sieved beforehand. The one sediment with a composition that most closely resembles the pastes of the preponderance of the pottery from the earthwork sites (no. 33-307 in Table 2.15) was recovered during NPS excavations at Hopeton at the base of wall segments at a depth of ca. one meter. This sediment is surely not unique to Hopeton, but probably recurs in various comparable settings where slack-water deposition had occurred within the Scioto alluvial valley. It is likely that this sediment was first encountered,

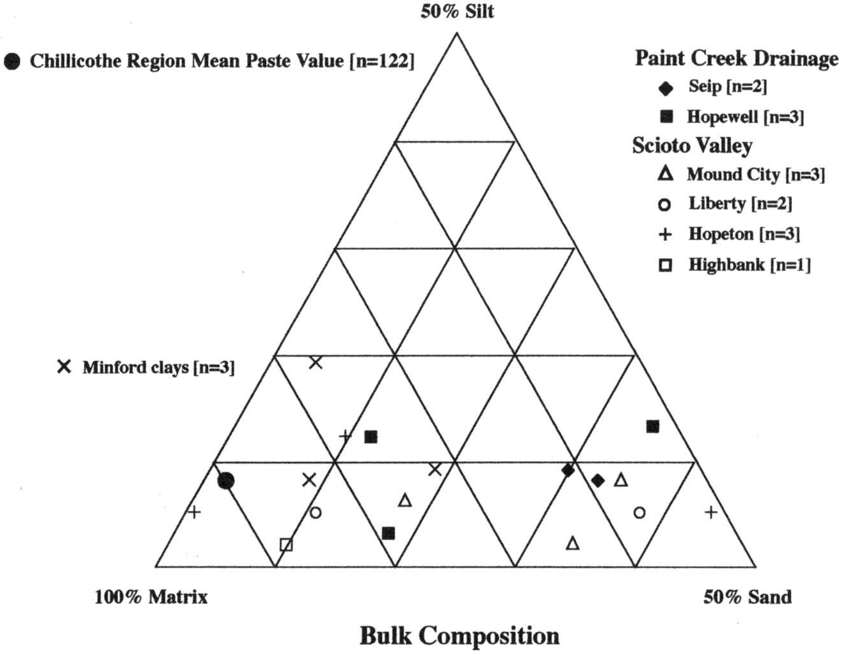

2.7. Ternary graph showing the paste values for 17 sediment samples from the Chillicothe region plotted against the paste mean for 122 local Hopewell vessels.

and its value quickly appreciated, as a result of the extensive earthmoving activities associated with the construction of the major Hopewell earthworks of the region. It is tempting to maintain that since no earthwork construction was conducted at the McGraw site, this deeply buried clay was unknown to that site's occupants.

Another possibility is that the vessels with higher silt values represent the utilization of Minford silts/clays. As can be seen from Table 2.15 and Figure 2.7, the three Minford samples available for this study are variable in composition, but all are relatively silty, i.e. have 8–19 percent silt. Two of these samples are notably sandier than local pastes, but one of them (no. 33-306, with 19 percent silt and 4 percent sand; Table 2.15; Figure 2.7) has a composition that is within the range of variation encountered among Hopewellian vessels at Seip (see Table 2.9) and Scioto vessels at McGraw (see Table 2.12). This raises the possibility that Hopewell potters may have exploited some settings within the broad geographic range of the Minford silts/clays. While leaving this possibility open, this study will proceed under the more parsimonious hypothesis that the silt variability observed within Chillicothe-region ceramics reflects the natural range of variation of immediately local alluvial clays.

An additional line of evidence may be cited in support of the view that the main clay sources used by Hopewell potters of the Chillicothe region were derived from riverine alluvium rather than from the lacustrine sediments of glacial Lake Tight. This pertains to the mineralogy of the sands in the clayey sediments in comparison to those observed in the pastes of local pottery. As documented in Table 2.15, grains of shale were observed in nine of the alluvial sediments, while seven of those same 13 sediments contained grains of limestone. Neither was recorded in the Minford silts/clays. Since both of these local rock types were frequently observed as natural inclusions in the local pottery, their presence points to the use of those alluvial sediments by Hopewell potters. Additionally, the occurrence of these rocks in pottery recovered outside of Ohio, especially shale (which is more geographically delimited than limestone), can be cited as supportive evidence for an Ohio origin.

The Turner Series: Local or Nonlocal?

The designation "Southeastern" for one of the Ohio Hopewell ceramic series clearly carries the implication of cultural interaction between the Southeastern United States and southern Ohio. That the formal properties inspiring this designation—simple stamping, check stamping, complicated stamping, and tetrapodal vessel bases—are common in Southeastern ceramics is undeniable (see Figure 2.3). Thus, the inference of cultural interaction between the two cultural provinces is unobjectionable. But, more specifically, what form did that cultural interaction take? Were pots being exchanged or merely ideas? If pots were being exchanged, were they handed down the line without population movements or did immigrants, traders, pilgrims, or other such mobile peoples transport them? Determining whether any of the Turner-series vessels were imported into Ohio cannot be definitively addressed on the basis of stylistic properties alone. Compositional evidence is essential.

Several vessels originally assigned to this series—those with sand and limestone tempers, which are common in Southeastern pottery—were set aside for later evaluation. This leaves twenty, grit-tempered Southeastern-series vessels unaccounted for. One of these vessels, represented by a plain tetrapod from Seip (see Table 2.11), was assigned directly to the Scioto series because of its lack of decoration and its seemingly local-in-every-way composition. The nonlocal status of the remaining 19 vessels, which are of the Turner Simple-stamped A (n=13) and Turner Check-stamped (n=6) types, are here evaluated because of their distinctively Southeastern style of decoration.

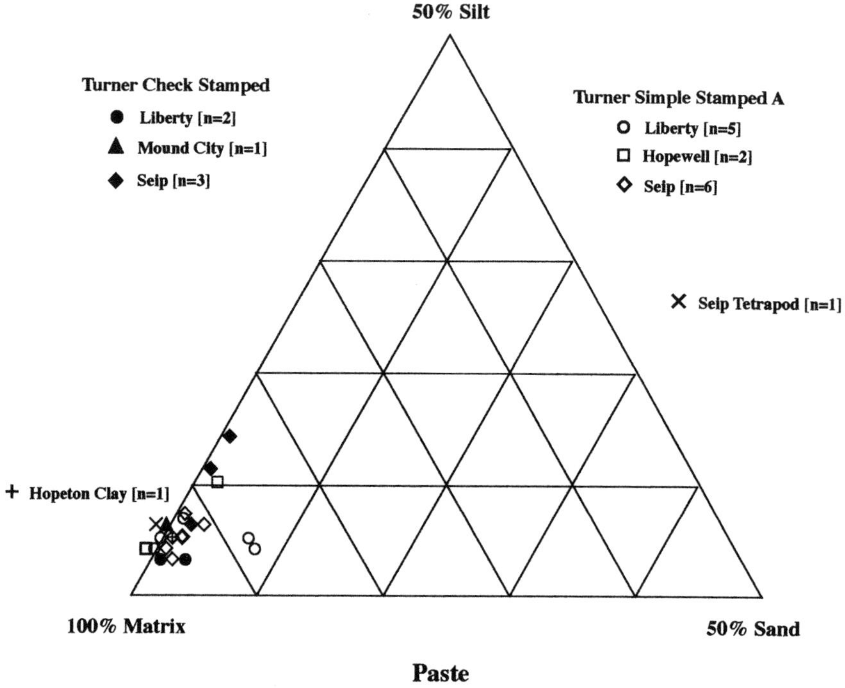

2.8. Ternary graph showing the paste values for each of the 19 Turner-series vessels from Ohio sites plus the tetrapodal base from Seip plotted against the sediment recovered from beneath the Hopeton Earthworks.

As can be seen from Tables 2.13 and 2.14, the mean compositional values for these Turner vessels differ in no significant ways from those of the local Hopewell and Scioto vessels at the same sites (see also Figure 2.6). One exception, vessel 33-78 from Mound City (see Table 2.5), has an unusually high incidence of temper, but this seems an insufficient basis for rejecting its local manufacture. The compositional evidence, then, supports the view that the majority of the Turner-series vessels were produced locally within the Chillicothe region (Figure 2.8).

Local Pottery from Eastern Ohio

The Marietta Earthworks

The Marietta site, one of the easternmost Hopewellian earthworks, is situated at the confluence of the Muskingum and Ohio rivers within the town of Marietta. The site is noteworthy because of the presence of no less than four flat-topped mounds within the larger of its two rectangular enclosures (Squier and Davis 1848:Plate 26). The southernmost of the platforms (Squier and Davis 1848:Plate 26b), known as the Capitolium Mound, has served as a substructure for the Washington County Public Library since 1915. Due to planned library expansion, the Cleveland Museum of Natural History conducted salvage excavations in the mound in 1990 (Pickard 1996). These excavations confirmed that the Marietta platforms were not of Mississippian affiliation as was still believed by some as late as 1978 (e.g., Brose 1978), but Hopewellian in origin (Pickard 1996).

Sherds from ten vessels recovered during the 1990 Cleveland Museum of Natural History excavations in the Capitolium mound (Pickard 1996), were thin sectioned for this study. The paste and body values for these vessels are recorded in Table 2.16. All are Scioto-series types, with eight grit tempered, one limestone tempered and one tempered with sandstone. No Hopewellian or Southeastern-series vessels are included.

With exceptions of incidental grog in one vessel and limestone and sandstone tempers in each of two others, the Marietta vessels in this study are characterized by a distinctive coarse-grained metamorphic rock of granitic composition, a metagranite (Figure 2.9). The minerals present in this rock include polycrystalline quartz accompanied by both plagioclase and microcline feldspars. The feldspars are commonly extensively altered and frequently possess perthitic intergrowth textures. Amphiboles and opaques also occur. An annealed texture and mylonitization provide unambiguous evidence of recrystallization.

With no samples available from local, clay-rich sediments, the main basis for characterizing the composition of locally made pottery is predicated on the assumption that the stylistically "local" pottery recovered at the site was made there. As can be seen from Table 2.16, all paste values fall well within the 95 percent confidence limits of the two-sigma standard for the site's eight local vessels. In addition all body values except the matrix percentage for one vessel (no. 33-291) fall within the site's two-sigma range (Table 2.16). Since that value is but .2 percent below the lower limit of 70.2 percent, sampling error due to the small sample size is the likely explanation. In light of the high degree of compositional homogeneity of the ten vessels, including the limestone and sandstone-tempered ones (sandstone is a common local bedrock), the assumption of local manufacture for all vessels is accepted as reasonable.

Knight Hollow Rockshelter

Knight Hollow is a multicomponent rockshelter located in the sandstone hills of Muskingum County in southeastern Ohio, some 50–60 km northwest of

Table 2.16. Body and Paste Values for Vessels from Two Sites in Muskingum County, Ohio, the Capitolium Mound at Marietta, and Knight Hollow Rockshelter

Value in bold falls outside the relevant ± two-sigma range shown in the shaded cell.

		Body					*Paste*			
Type	**Thin Section No.**	**% Matrix**	**% Sand**	**% Temper**	**Temper Size Index**	**Temper Type**	**% Matrix**	**% Silt**	**% Sand**	**Sand Size Index**
Capitolium Mound										
McGraw Plain	33-290	81	1	18	3.88	Granite	95	4	1	1.00
McGraw Cordmarked	33-291	**70**	5	25	4.20	Granite	88	6	6	3.33
McGraw Cordmarked	33-292	85	4	11	3.40	Granite	85	11	4	1.71
McGraw Cordmarked	33-295	80	2	18	4.11	Granite	89	8	3	2.50
McGraw Cordmarked	33-296	84	6	10	3.50	Granite	90	3	7	3.00
McGraw Cordmarked	33-297	86	1	13[10]	3.54	Granite+	95	4	1	1.00
				[3]	3.67	Grog				
McGraw Cordmarked	33-298	80	0	20	4.00	Granite	96	4	0	—
McGraw Cordmarked	33-299	91	2	7	3.88	Granite	87	11	2	1.83
Mean ±1 Std Deviation	**[n=8]**	82.0±5.9	2.8±2.1	15.2±6.0	3.92±.43		90.5±4.3	6.4±3.2	3.1±2.5	2.05±.92
Two-Sigma Ranges	**[n=8]**	70.2–93.8	0–7.0	3.2–27.2	3.06–4.78		81.8–99.1	0–12.8	0–8.1	1.00–3.89
Possibly Nonlocal										
McGraw Cordmarked	33-293	83	4	13	3.62	Limestone	88	8	4	2.00
McGraw Cordmarked	33-294	76	2	22	3.88	Sandstone	87	10	3	1.67
Knight Hollow										
Turner Simple St-A	33-300	84	0	16	3.24	Granite	97	3	0	—

2.9. Photomicrograph of metagranite temper in local Vessel 33-291 from Marietta taken at 10X magnification under crossed polars. Scale: grain at top=2.15 mm. The two largest grains show evidence of mechanical deformation (the areas of extremely fine-grained quartz) referred to as mylonitization [My in figure]. Plagioclase feldspar (with twinning) also present.

Marietta. Three body sherds and a large rim from a simple-stamped vessel were recovered here in 1964 (Felumlee 1983:23 and Fig. 13). With the encouragement of N'omi Greber, James Morton obtained and sent to me one of the body sherds for thin sectioning. The composition of this thin section (33-300) is recorded in Table 2.16. As can be seen from this table, this vessel is tempered with granite and has compositional values that fall fully within the two-sigma range of the eight Marietta vessels. There can be no reasonable doubt that this is a local product of southeastern Ohio.

The Newark Earthworks

The Newark Earthworks are located in Liberty County in central Ohio in the headwaters of the Muskingum River, which enters the Ohio River at Marietta. The largest and most complex of Ohio's Hopewellian earthworks, the site has seen little professional investigation (Lepper 1988; see also Squier and Davis 1848:Plate 25).

The four sherds included in this study were recovered during salvage excavations from 1977 to 1980 within a highway right-of-way that traversed the city of Newark while also encroaching upon portions of the earthworks

Table 2.17. Body and Paste Values by Vessel from 33LI252 Near the Newark Earthworks

All compositional values fall within the relevant ± two standard deviations/95% confidence limits shown in the last rows.

Paste

Type	Thin Section No.	% Matrix	% Silt	% Sand	Sand Size Index
McGraw Cordmarked	33-234	85	12	3	1.00
McGraw Cordmarked	33-235	81	14	5	1.00
Mean ±1 Std Deviation	**[n=2]**	83.0±2.8	13.0±1.4	4.0±1.4	1.00±0
Two-Sigma Ranges	**[n=2]**	77.4-88.6	10.2-15.8	1.2-6.8	1.00

Body

Type	Thin Section #	% Matrix	% Sand	% Temper	Temper Size Index	Temper Type
McGraw Cordmarked	33-234	83	3	14	3.20	Granite
McGraw Cordmarked	33-235	88	4	8	3.25	Granite
Mean ±1 Std Deviation	**[n=2]**	85.5±3.5	3.5±0.7	11.0±4.2	3.22±.04	
Two-Sigma Ranges	**[n=2]**	78.5-92.5	2.1-4.9	2.6-19.4	3.14-3.30	

themselves (Lepper and Yerkes 1997). The sherds were recovered from two features within what was apparently a seasonal, small-group residential site—33LI252—located within 600 meters of the northeast end of the earthworks (Lepper and Yerkes 1997:190–193). Two of the sherds, one each from Features 3 and 15, are simple stamped and suspected on both stylistic and compositional grounds of being nonlocal in origin. The other two sherds, both from Feature 15, are from local McGraw Cordmarked vessels. The compositional data for the two local vessels are presented in Table 2.17.

The two cordmarked vessels are tempered with a distinctive metagranite whose composition is dominated by untwinned, heavily altered feldspars with quartz in a distinct minority. Compared to Scioto-series vessels elsewhere in Ohio, the volume and size of temper in these Newark vessels is on the low end of the range of variation, but not abnormally so (cf. Tables 2.13 and 2.17). By contrast, the paste indices show relatively high amounts of silt, comparable to those of Scioto-series vessels at the McGraw site (see Table

2.12). These observations made, however, the problem of sample inadequacy cannot be overestimated in trying to characterize Newark-area, Scioto-series pottery compositions. These data are useful to the extent that they can be contrasted with the two simple-stamped vessels also recovered here. We shall return to this topic in Chapter 10 after potential Southeastern sources for the latter two vessels are considered.

Local Pottery from Southern Ohio: the Tremper Site

W. C. Mills excavated the Tremper mound, located in the lower Scioto river valley just north of Portsmouth, in 1915 (Mills 1916). The site is well known for the large number of platform pipes, over 140 in number, that Mills recovered here (Mills 1916:127). In his analysis of the surviving pottery from the site in the collections of the Ohio State Historical Society, Prufer (1968:45–50) tabulated a total of 431 sherds, 420 attributed to the Scioto series, six to the Southeastern series, and five untyped (Prufer 1968:46). Mention was also made of "About half-a-dozen" Seip Plain (i.e. polished but undecorated) sherds, but since none was a rim, they were included in the Scioto-series tabulations (Prufer 1968:46). Two aspects of this collection merit emphasis: (1) no decorated Hopewellian-series vessels are present and (2) the untyped vessels, all of which were incised, are characterized as "clay tempered" and considered "completely atypical of any known Ohio Hopewell affiliated ceramic series" (Prufer 1968:46). This is the only report of which I am aware that mentions the presence of grog temper in Ohio Hopewell contexts.

For this study, sherds from nine vessels from the Tremper site collections at the Ohio State Historical Society were thin sectioned. Four are McGraw Plain, three Seip Plain, and two Turner Simple-stamped A (Table 2.18). As can be seen from Table 2.18, the paste values for these vessels are relatively uniform across the three major pottery series. All vessels are comprised of >90 percent matrix, <9 percent silt, and <4 percent sand. Because of this paste homogeneity, the values for vessels from all three series are summed to provide a grand paste mean for the site. Using this grand mean to provide two-sigma confidence intervals for the site, it can be seen that seven of the nine vessels have values that fall within these ranges (see Table 2.18). Two vessels have single values that fall beyond the two-sigma limits, both by less than one percentage point (vessels 33-214 and 33-219 in Table 2.18). Considering the small sherd sample from Tremper, it is likely that these values reflect nothing more than sampling error, i.e., the thin-sectioned vessels do not fully represent the actual compositional range of variation at the site.

While the Tremper paste values are similar to those of most other Ohio Hopewell sites, the body properties are unique within that context. Grog, i.e.,

Table 2.18. Body and Paste Values by Vessel for Tremper Mound

Values in bold fall outside the relevant ± two standard deviation ranges shown in the shaded cells.

		Body					*Paste*			
Type	**Thin Section No.**	**% Matrix**	**% Sand**	**% Temper**	**Temper Size Index**	**Temper Type**	**% Matrix**	**% Silt**	**% Sand**	**Sand Size Index**
Scioto Series										
McGraw Plain	33-214	85	0	15[14+1]	3.75/4.00	Grog+Grit	92	**8**	0	—
McGraw Plain	33-215	75	1	24[23+1]	4.50/5.00	Grog+Grit	95	3	2	1.00
McGraw Plain	33-218	70	0	30	4.07	Grog	94	6	0	—
McGraw Plain	33-219	67	**3**	30	4.00	Grog	92	4	**4**	1.00
Mean ±1 Std Dev	**[n=4]**	74.25±7.9	1.0±1.4	24.75±7.1 Gg-24.25±7.6 Gt-.50±0.6	4.02±..36 Gg-4.08±.31 Gt-4.50±.70		93.25±1.5	5.25±2.2	1.5±1.9	1.00±.0
Hopewellian Series										
Seip Plain	33-213	79	0	21[12+9]	4.00/3.62	Grog+Grit	97	3	0	—
Seip Plain	33-216	80	0	20[13+7]	4.12/3.75	Grog+Grit	97	3	0	—
Seip Plain	33-217	95	0	5[4+1]	3.80/3.00	Grog+Grit	95	5	0	—
Mean ±1 Std Dev	**[n=3]**	84.7±9.0	0	15.3±9.0 Gg-9.7±4.9 Gt-5.6±4.2	3.79±.22 Gg-3.97±.16 Gt-3.46±.40		96.3±1.2	3.7±1.2	0	—
Turner Series										
Turner Simple-St A	33-211	88	0	12[11+1]	3.80/4.00	Grog+Grit	97	3	0	—

Turner Simple-St A	33-212	97	0	3	3.40	Grog	98	2	0	—
Mean ±1 Std Dev	**[n=2]**	92.5±6.4	0	7.5±6.4 Gg-7.0±5.7 Gt-0.5±0.7	3.73±.31 Gg-3.60±.30 Gt-4,00		97.5±0.7	2.5±0.7	0	—
Mean ±1 Std Dev	**[n=9]**	81.8±10.4	0.4±1.0	17.8±9.8 Gg-15.6±10.0 Gt-2.2±3.3	3.85±.32 Gg-3.94±.30 Gt-3.90±.65		95.2±2.2	4.1±1.9	0.7±1.4	1.00±0
Two-Sigma Range	**[n=9]**	61.00-100	0-2.4	Gg-0-35.6	Gg-3.34-4.54		90.9-99.6	0.3-7.9	0-3.5	1.00

Note: St=Stamped; Std=Standard; Std Dev=Standard Deviation; Gg=Grog; Gt=Grit

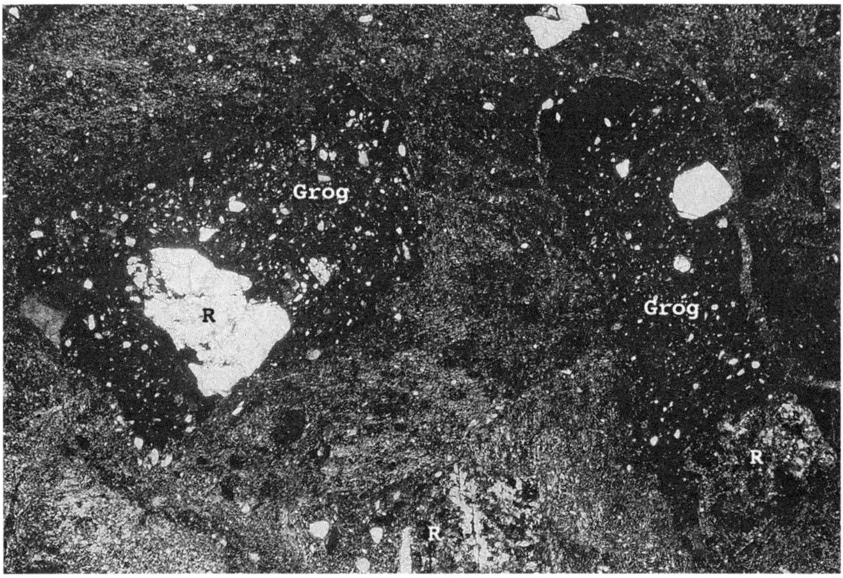

2.10. Photomicrograph of grit-tempered grog in local Vessel 33-211 from Tremper taken at 10X magnification under crossed polars. Two grains of grog, each with rock inclusions, are on opposite sides of the field of view with a rock fragment partially visible at the bottom. Three rock grains are denoted by R. Scale: long grog on right measures 2.25 mm.

fragmented pottery, is the predominant temper in all nine vessels, ranging in value from 3% to 30% by volume (see Table 2.18). Coarse, angular fragments of a variety of igneous rocks, including granite, diorite, and gabbro, were also observed and tabulated as "grit." Several of these rock fragments were probably derived from the grog, for in at least four instances grit-tempered grog was observed (Figure 2.10). There was also at least one occurrence of grog-tempered grog. Interestingly, despite the temper differences between Tremper and the Chillicothe-region sites, the practice of using lesser amounts and finer sizes of temper in the Hopewellian series (accepting the Seip Plain vessels as in this series) in contrast to the Scioto series was also apparently in vogue at Tremper (see Table 2.18)

If the current sample of nine thin-sectioned vessels is representative, grog was the prevalent, if not the sole, temper at the site. As noted previously, Prufer (1968:46) mentions the presence of "clay tempered" sherds, but this observation applied only to five untyped, incised sherds, none of which was included in the thin section sample. At least for the present, Tremper stands out as unique among Ohio Hopewell sites in being characterized primarily by grog-tempered pottery.

Local Pottery from Southwestern Ohio: the Turner Site

The Turner site is located near the confluence of the Little Miami and Ohio Rivers in the southwest corner of Ohio (see Figure 2.1). The Peabody Museum of Harvard University, under the overall direction of Frederick Ward Putnam, conducted substantial excavations here between 1882 and 1908 (Willoughby and Hooten 1922). The 36 Turner-site sherds thin sectioned for this study were obtained from the Peabody Museum collections. Table 2.1 records the relative frequencies of sherds by ceramic series for the site.

On stylistic and compositional grounds, 15 of the 36 vessels are regarded as probable imports from the Southeast, so their consideration will be deferred to Chapter 11. Eight of the 36 vessels are limestone tempered and will be discussed at the end of this chapter along with the other limestone-tempered vessels from Ohio. The remaining 13 vessels are grit tempered and stylistically local in the sense that they are cordmarked (n=3), plain (n=4), or have distinctive Turner series (n=2) or Hopewellian series (n=4) characteristics. The individual compositional values for these 13 vessels are recorded in Table 2.19.

Since no local sediment samples are available, the paste values for these 13 vessels provide the best available basis for estimating the composition of locally made pottery. As can be seen from Table 2.19, the paste values for all vessels fall within the two-sigma confidence limits defined for their respective series. Interestingly, one vessel (no. 33-272) has a .3 percent difference from the two-sigma paste range for silt for all 13 vessels yet falls within the two-sigma range for the seven Scioto-series vessels from the site. This minor anomaly can certainly be attributed to sampling error.

As with other Ohio Hopewell sites in this study, there appear to be notable differences in body composition between Hopewellian-series and Scioto-series vessels at Turner. The temper size indices for the two series are closely similar, but differences in temper percentages are consistent with the pattern previously observed; namely, the Scioto series has more temper—21 percent vs. 16 percent—than the Hopewellian series (see Table 2.19).

Limestone-Tempered Pottery in Hopewell Sites of Southern Ohio

Let us now turn to the issue of the local manufacture of limestone-tempered pottery at the Hopewell sites of southern Ohio. At the outset of this discussion it is important to note that limestone bedrock occurs widely in southern Ohio (e.g., Lamborn et al. 1938). Prufer (1968:7–15) included limestone-tempered vessels in all three of his local ceramic series. However, because of the widespread occurrence of limestone temper in ceramic assemblages

Table 2.19. Body and Paste Values for the Grit-Tempered Vessels by Ceramic Series at the Turner Site

Value in bold exceeds the ± two-sigma range shown in the shaded cell.

		Body					***Paste***			
Type	**Thin Section No.**	**% Matrix**	**% Sand**	**% Temper**	**Temper Size Index**	**Temper Type**	**% Matrix**	**% Silt**	**% Sand**	**Sand Size Index**
Hopewellian Series										
Rocker Stamped	33-266	82	0	18	3.73	Granite	95	5	0	—
Rocker Stamped	33-279	89	1	10	4.00	Granite	94	5	1	3.00
Seip Plain	33-288	78	4	18[11+7]	3.27	Granite+ Gabbro	91	4	5	5.00
Seip Plain	33-289	82	0	18	3.68	Granite	99	1	0	—
Mean ±1 Std Deviation	**[n=4]**	82.8±4.6	1.2±1.9	16.0±4.0	3.67±.30		94.8±3.3	3.7±1.9	1.5±2.4	4.00±1.41
Two-Sigma Ranges	**[n=4]**	73.6-92.0	0-5	8.0-24.0	3.07-4.27		88.2-100	0-7.5	0-6.3	1.18-6.82
Scioto Series										
McGraw Cordmarked	33-256	72	8	20	4.54	Diorite	85	6	9	1.58
McGraw Cordmarked	33-260	72	9	19[17+2]	3.53/4.00	Granite+ Grog	81	8	11	2.00
McGraw Cordmarked	33-261	77	8	15[14+1]	3.09	Granite+ Gabbro	82	8	10	1.56
McGraw Plain	33-269	71	3	26	3.74	Granite	90	7	3	1.60
McGraw Plain	33-271	**83**	0	17	3.11	Granite	94	6	0	—
McGraw Plain	33-272	73	3	24	4.04	Granite	84	12	4	2.00
McGraw Plain	33-273	73	0	27[22+5]	3.39	Granite+ Gabbro	95	5	0	—

Mean ±1 Std Deviation	**[n=7]**	74.4±4.2	4.4±3.9	21.2±4.6	3.63±.52		87.3±5.8	7.4±2.3	5.3±4.7	1.75±.23
Two-Sigma Ranges	**[n=7]**	66.0-82.8	0-12.2	12.0-30.4	2.59-4.67		75.7-98.9	2.8-12.0	0-14.7	1.29-2.21
Turner Series										
Simple-Stamped A	33-265	84	1	15	2.66	Granite	93	6	1	1.00
Check-Stamped	33-280	57	9	34[26+8]	3.57	Granite+ Gabbro	79	9	12	1.58
Mean ±1 Std Deviation	[n=2]	70.5±19.1	5.0±5.7	24.5±13.4	3.12±.64		86.0±9.9	7.5±2.1	6.5±7.8	1.29±.41
Two-Sigma Ranges	**[n=2]**	32.3-100	0-16.4	0-51.3	1.84-4.40		66.2-100	3.3-11.7	0-22.1	1.00-2.11
Site Mean ±1 Std Deviation	**[n=13]**	76.4±8.1	3.5±3.6	20.1±6.2	3.56±.48		89.4±6.4	6.3±2.7	4.3±4.6	2.15±1.20
Site Two-Sigma Ranges	**[n=13]**	60.2-92.6	1-10.7	7.7-32.6	2.60-4.52		76.6-100	0.9-11.7	0-13.5	1.00-4.55

Note: Mean temper size pertains to grit only.

of the Midsouth, such vessels were initially relegated to the "Possibly Nonlocal" category until their local manufacture could be independently investigated through compositional analysis. Now that all the compositional data for the limestone-tempered vessels in our Ohio sample have been compiled, let us consider them as a group.

As recorded in Table 2.1, 18 such vessels from 5 Ohio sites were thin sectioned for this study. Because the physical properties of limestones are so different from those of igneous rocks (see Figure 2.5), comparing body indices for vessels possessing these tempers is problematic, because temper size and amount are primary determinants of body values. For this reason, the focus in reviewing the compositions of the 18 limestone-tempered vessels will be upon the pastes, which are determined by excluding temper from consideration.

Table 2.20 records the paste values by site for each of the 18 limestone-tempered vessels from southern Ohio along with the two-sigma paste ranges for the local vessels from each site. By comparing the paste values for each vessel with the two-sigma ranges for the appropriate site, it can be seen from Table 2.20 that five of the 18 vessels (no. 33-99 from Mound City, nos. 33-24 and 33-249 from Liberty, and nos. 33-257 and 33-275 from Turner) have paste values that depart from the two-sigma ranges of the local vessels of their respective sites. The anomalous values for vessels 33-99, 33-240, 33-249, and 33-257 all depart by less than 2 percentage points from their respective two-sigma ranges/95 percent confidence limits, rendering it most likely that these values are artifacts of the small samples involved. McGraw Plain vessel 33-275 from the Turner site, on the other hand, has a notably sandy paste suggesting that it was made from a distinctive, perhaps nonlocal, clay. The remaining 13 vessels in Table 2.20 have paste values that fall fully within the two-sigma ranges of the local vessels at their respective sites. In sum, it appears that most of the limestone-tempered vessels recovered in Ohio Hopewell sites that have been thin sectioned can best be regarded as manufactured from local clays, i.e., were not imported from elsewhere. This issue will be reconsidered in Chapter 8 when both body as well as paste values for limestone-tempered pottery from several Southeastern sites are analyzed.

Overview: Hopewell Pottery of Southern Ohio

Petrographic analysis has led to several new insights concerning the composition, manufacture, and use of Ohio Hopewell pottery. At the most basic level, those raw materials used in pottery manufacture (i.e., tempers and pastes) have been identified, and the mix of those raw materials (i.e., body) used in

Table 2.20. Paste Values for All [n=18] Limestone-Tempered Vessels from Ohio vs. Local Paste Two-Sigma Ranges for Each Site

Values in bold fall outside the relevant ± two-sigma ranges shown in shaded cells.

	Thin Section No.	% Matrix	% Silt	% Sand	Sand Size Index
Mound City					
Turner Check-Stamped	33-79	97	2	1	1.00
McGraw Cordmarked	33-99	82	**17**	1	1.00
Mound City 2-Sigma Ranges	(n=26)	79.6-99.2	1.6–15.2	0–9.2	1.00–1.77
Liberty					
McGraw CM (RB-Md 2)	33-240	**82**	9	**9**	1.71
McGraw CM (RB-Md 2)	33-249	85	**10**	5	1.89
Turner S-s A (Harness)	33-131	94	1	5	1.25
Turner S-s A (Harness)	33-144	92	1	7	1.22
Liberty 2-Sigma Ranges	(n=38)	84.7–99.9	0.1–9.7	0–8.6	1.00–2.67
Seip					
McGraw Plain	33-123	95	4	1	1.00
McGraw Plain	33-124	99	1	0	—
Turner Check-Stamped	33-198	96	4	0	—
Seip 2-Sigma Ranges	(n=39)	78.7–99.5	0–19.2	0–5.1	1.00–2.12
Marietta					
McGraw Cordmarked	33-293	88	8	4	2.00
Marietta 2-Sigma Ranges	(n=8)	81.8–99.1	0–12.8	0–8.1	1.00–3.39
Turner					
Hopewell Rim	33-276	97	3	0	—
Chillicothe Rocker-Stamped	33-278	90	9	1	1.50
McGraw Cordmarked	33-257	100	**0**	0	—
McGraw Cordmarked	33-258	89	6	5	1.67
McGraw Cordmarked	33-259	93	6	1	2.50
McGraw Plain	33-270	87	7	6	1.27
McGraw Plain	33-274	94	6	0	—
McGraw Plain	33-275	**71**	10	**19**	1.91
Turner 2-Sigma Ranges	(n=13)	76.6–100	0.9–11.7	0–13.5	1.00–4.55

the manufacture of pottery vessels has been determined. These data provide the foundation for identifying regional differences in ceramic composition, the use of different recipes for different kinds of vessels, for objectively discriminating local from nonlocal pottery, and for evaluating the status of possible exports from Ohio recovered elsewhere in the Eastern United States.

An unanticipated finding was evidence that Ohio Hopewell potters used significantly different recipes in the manufacture of Scioto-series as opposed to Hopewell-series vessels. At all sites with sufficient samples of both series, a similar pattern was observed; namely, the volume of temper within the Scioto series is greater than in the Hopewell series. In addition, with but one exception (the Turner site), the temper sizes are greater in Scioto-series vessels. The inference drawn from these data is that Hopewell potters of Ohio intentionally employed different recipes in the manufacture of their pottery vessels in order to maximize the effectiveness of those vessels in performing the kinds of tasks envisioned for them.

Another finding, perhaps anticipated but nonetheless useful, involved confirmation that many vessels of Prufer's Southeastern series were made of the same raw materials as demonstrably local pottery. In other words, the compositional evidence confirms that many vessels with simple and check stamping, decorative treatments widely found in the Southeast, were locally produced in southern Ohio. In a similar vein most of the thin-sectioned vessels with limestone temper, a common temper medium in the mid-continent south of the Ohio River, also were apparently made from local clays. This, of course, does not mean that all such vessels recovered in Ohio Hopewell contexts were locally made, but that *many* such vessels were.

Based upon the compositional data, some interesting regional differences in pottery making practices were documented within Ohio Hopewell society. Most notable was the discovery that grog was the prevalent temper at the Tremper site. This is not so surprising considering that, as will be discussed presently, the use of grog temper was common downstream along the Ohio River in Indiana and Illinois in Middle Woodland contexts. It makes sense to view this practice, as is also the case in the Lower Mississippi Valley, as an adaptation to a large alluvial valley that is generally deficient in suitable rock tempers.

Beside the predominance of granitic rock tempers, other compositional properties of Ohio Hopewell pottery that can be of value in identifying vessels exported from this region include the paste characteristics. Table 2.21 provides a summary tabulation of regional paste indices for the Hopewell sites of southern Ohio. As a generalization, the pastes of Ohio Hopewell vessels are typically comprised of ca. 89–96 percent matrix, with the combined to-

Table 2.21. Summary of Ohio Hopewell Paste Values (Means and Standard Deviations) by Site

Site	N	% Matrix	% Silt	% Sand	Sand Size Index
Liberty	38	92.3±3.8	4.9±2.4	2.8±2.9	1.59±.54
Mound City	26	89.4±4.9	8.4±3.4	2.2±3.5	1.21±.28
Hopewell	13	89.5±3.8	7.6±2.4	2.9±2.2	1.98±.91
Seip	39	89.1±5.2	9.2±5.0	1.7±1.7	1.22±.45
McGraw	6	84.3±6.5	14.2±5.3	1.5±1.9	1.72±.91
Chillicothe Region	122	90.0±5.0	7.7±4.4	2.3±2.6	1.44±.59
Marietta	8	90.5±4.3	6.4±3.2	3.1±2.5	2.05±.92
Newark (33LI252)	2	83.0±2.8	13.0±1.4	4.0±1.4	1.00±0
Tremper	9	95.2±2.2	4.1±1.9	0.7±1.4	1.00±0
Turner	13	89.4±6.4	6.3±2.7	4.3±4.6	2.15±1.20

tals of silt and sand commonly less than 10 percent and rarely greater than 15 percent; in the vast majority of cases the silt percentages exceed those of sand. The sand size indices rarely exceed 2.00, and when they do, the reason is usually distinctive grains of shale, sandstone, or limestone, all of which are common minor, natural inclusions in Ohio Hopewell pottery. Vessels recovered outside of Ohio that have such paste values along with granitic temper can be accepted as credible candidates for intrusives from Ohio.

The paste values of ceramics from the Hopewell sites of Ohio are surprisingly uniform as a group (see Table 2.21), which creates a problem for efforts to identify specific sites based upon composition. As previously discussed, several vessels with high silt values (i.e., >10 percent) are unusual enough in composition to be considered as possibly nonlocal. For example, it is interesting to note that silt values of 11–21 percent (see Table 2.10) characterize all 14 of the Hopewell-series vessels from Seip, but only 6 of the remaining 25 local vessels from the site have silt values greater than 7.5 percent (see Tables 2.11 and 2.12). The two sites with the smallest samples, McGraw (n=6) and Newark (n=2), also have distinctly high-silt pastes (see Table 2.21), which raises the red flag of sampling error. Because at least one or two such high-silt vessels occur at literally every Ohio site sampled in this study, the view that these vessels were of nonlocal origin is further rendered implausible. In conclusion, it can be suggested that the compositional evidence provided by this thin section study should permit the reasonably reliable identification of pottery vessels exported from Ohio when both body and paste indices are considered, but tracing such vessels to a specific Ohio site, excepting the unique Tremper site, is problematic at best.

Ohio Hopewell Pottery Recovered Elsewhere in the Eastern Woodlands

With discussion of the composition of Ohio Hopewell pottery just concluded, this seems the optimal place to consider the topic of possible Ohio exports elsewhere in the Eastern Woodlands. Despite the extensive evidence of cultural interaction between the Hopewell centers of Ohio and Middle Woodland peoples across the East (e.g., Seeman 1979), the export of pottery vessels from Ohio was apparently an extraordinarily rare occurrence. In evaluating the possibility that pottery vessels recovered outside of Ohio were imported from Ohio, the strongest cases for importation can be put forward when two criteria are met: (1) the suspected exotic vessels are stylistically anomalous while simultaneously resembling Ohio Hopewell pottery; and (2) the compositions (i.e., paste and body parameters) are unlike local pottery but similar to Ohio Hopewell pottery. The first criterion is neither necessary nor sufficient to provide a convincing case for importation, because local potters can readily adopt exotic stylistic properties through various mechanisms of cultural interaction not involving the physical transmission of pottery vessels. But, in combination with the compositional evidence, the significance of the stylistic properties is certainly enhanced.

So far, within the context of the present study, a strong case can be made for the occurrence of Ohio Hopewell pottery in a Middle Woodland setting outside of Ohio only for two sites. Those sites are Icehouse Bottom in southeastern Tennessee (Stoltman 1999b) and Mann in southwestern Indiana (Figure 3.1).

Turning first to Icehouse Bottom, 92 percent of the site's pottery was assigned to two local series, Candy Creek, which is limestone tempered, and Connestee, which is sand tempered (Chapman 1973:41). In addition 21 sherds, constituting just .1 percent of the site's pottery (Chapman 1973:41), were assigned to a "Hopewell Series" based upon incised and rocker-stamped decoration closely resembling Ohio Hopewell motifs (Chapman 1973:58–60). Both grit and limestone tempers were observed among vessels of the Hopewell series (Chapman 1973:58–60). Jefferson Chapman of the University of Tennessee provided sherds from three of these latter vessels for thin section analysis, all characterized by rocker-stamped decoration (Figure 2.11). One of them (no. 40-8) also has a trailed line. Two of these sherds (nos. 40-6

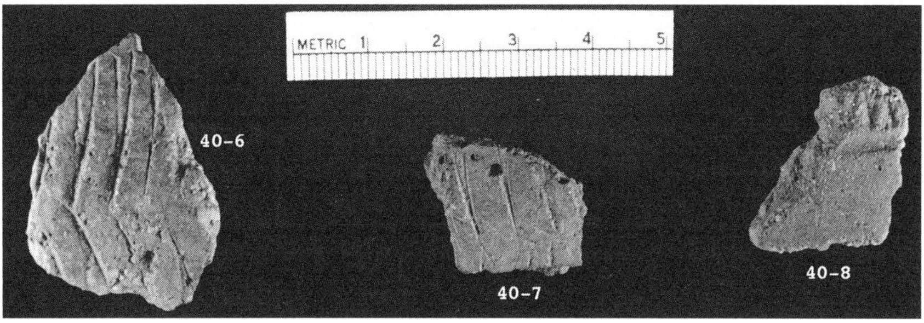

2.11. Three rocker-stamped sherds from Icehouse Bottom, each associated with its thin section number. Nos. 40-6 and 40-8 are believed to be imports from Ohio, while no. 40-7 is probably of local manufacture.

2.12. Photomicrograph of granite-tempered vessel 40–8 from Icehouse Bottom taken at 10X magnification under crossed polars. Scale: large granite grain (lower left)=1.15mm with microcline (Mi), quartz (Q), and altered feldspar (F); shale (upper right)= .975 mm. This vessel was imported to Icehouse Bottom from Ohio.

and 40-8) were determined to have granitic rock temper (e.g., Figure 2.12), whereas the third (no. 40-7) was sand tempered and is probably a local product (see Chapter 5 for further discussion).

Table 2.22 records the compositional data for the two grit-tempered vessels, which conform closely to the two criteria set forth above; namely, (1) they are stylistically anomalous in southeastern Tennessee while being indis-

Table 2.22. Paste and Body Values for Three Grit-Tempered Vessels Recovered Outside Ohio Considered Probable Imports from Ohio Compared with Mean Hopewellian-Series Values for the Chillicothe Region

The value in bold exceeds the relevant ± two-sigma range shown in the shaded cell.

Paste

Type	Thin Section No.	% Matrix	% Silt	% Sand	Sand Size Index
Icehouse Bottom, TN					
Plain Rocker Stamped	40-6	91	7	2	1.57
Zoned Rocker Stamped	40-8	89	6	5	1.57
Mann, IN					
Plain	42	90	8	2	1.00
Chillicothe-Region Mean ±1 Standard Deviation	[n=122]	90.0±5.0	7.7±4.4	2.3±2.6	1.44±.59
Chillicothe-Region 2-Sigma Ranges	[n=122]	80.0–100	0–16.5	0–7.5	1.00–2.62

Body

Type	Thin Section #	% Matrix	% Sand	% Temper	Temper Size Index
Icehouse Bottom, TN					
Plain Rocker Stamped	40-6	77	2	21	3.14
Zoned Rocker Stamped	40-8	78	**4**	18	2.38
Mann, IN					
Plain	42	78	2	20	2.60
Chillicothe-Region Mean					
±1 Standard Deviation (Hopewellian Series Only)	[n=37]	81.7±4.3	1.5±1.2	16.8±4.2	3.05±.47
Chillicothe-region 2-Sigma Ranges	[n=37]	73.1–90.3	0–3.9	8.8–25.2	2.11–3.99

tinguishable from Hopewellian Series pottery of Ohio, and (2) they are tempered with coarse grains of granite, just like Ohio Hopewell pottery, and have pastes that are compositionally fully within the range of variation of southern Ohio pottery. These two sherds, which presumably derive from different vessels, are here accepted as convincing candidates for Ohio Hopewell imports into Tennessee. Additional support for this view occurs in the form of shale grains that were observed in one of the vessels, no. 40-8 (see Figure 2.12).

Turning to the Mann site, one vessel (no. 42) from among the 53 thin-sectioned vessels that were available for this study appears to have been imported from Ohio. While bearing no decoration (although it could be from an undecorated zone of a decorated vessel), it is unique in having grit temper of syenitoid composition (i.e., the two feldspars, orthoclase and plagioclase, predominate while only minor amounts of quartz occur; see Le Maitre 1989). The compositional values shown in Table 2.22 reveal that this vessel conforms in every way to the properties of Chillicothe-region pottery. It, too, seems an excellent candidate for acceptance as an import from southern Ohio. While it is felt that the compositional data for all three of these vessels provide solid evidence for derivation from southern Ohio, it is currently impossible to identify specific site sources.

3

The Mann Site in Posey County, Indiana

In the following seven chapters we now turn to a presentation of the thin section data from sites outside of Ohio for which evidence of Hopewell interaction exists. Figure 3.1 shows the locations of these sites.

The Mann site is located near the confluence of the Wabash and Ohio Rivers in the southwestern corner of Indiana (see Figure 3.1). As with the major Hopewell centers of southern Ohio, an impressive complex of earthworks, including mounds and walled enclosures, characterize the Mann site. Unlike the Ohio sites, however, its surface is littered with a rich array of artifacts indicative of extensive, and perhaps also intensive, habitation by Middle Woodland peoples. Despite its enormous size—covering an area of at least 175 ha (Kellar 1979:100)—it has received limited professional archaeological investigation. The site first appeared on the radar screen of archaeologists in 1931 in the aftermath of a visit by Warren K. Moorehead and Glenn Black, among others, to Evansville College, where a substantial number of artifacts surface collected at Mann were housed (Adams 1949:47). It was not until three decades later that the first carefully controlled excavations were conducted at the site. James Kellar supervised these excavations in a series of archaeological field schools in behalf of Indiana University in 1964, 1966, 1967, and 1977. The most extensive report on the site occurs in a doctoral dissertation by Bret J. Ruby (1997), with the most readily accessible sources of information being two published papers (Kellar 1979; Ruby 2006).

For this study, thin sections from 54 pottery vessels and 7 sediment samples from the University of Indiana collections were analyzed. Christine Shriner, who had previously conducted her own analysis of the pottery thin sections, generously provided them ready-made (Ruby and Shriner 2005). As will be seen in the discussion to follow, the two petrographic analyses produced closely similar results, with the main difference, in line with the goals of this study, being the rendering of quantitative data from the Mann site

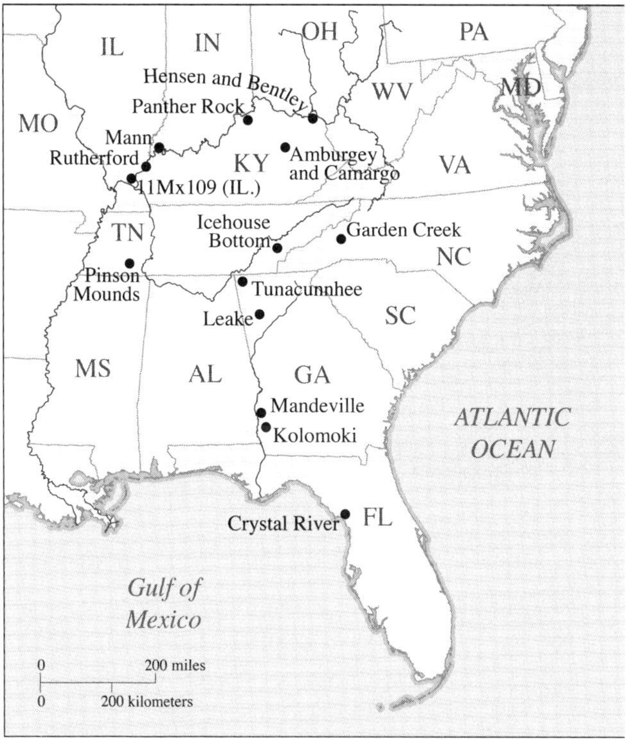

3.1. Map showing the locations of sites outside of Ohio for which thin section analyses were conducted in this study.

into a form that could be compared directly to the larger Middle Woodland database utilized in this study.

Kellar (1979:103) introduced what has become the standard typology for Mann ceramics (see also Ruby 1997 and Ruby and Shriner 2006) and is the typology used in this study. Patterned after Prufer's 1968 typology of Ohio Hopewell pottery, Kellar (1979103) recognized three basic classes of Mann pottery: Utilitarian, Southeastern, and Hopewellian. The former, analogous to Prufer's Scioto Series, is comprised of cordmarked and plain vessels. The Southeastern class was recognized on the basis of simple, check, and complicated stamped surfaces along with vessels with tetrapodal bases. The Hopewellian class was defined around vessels bearing incised, trailed, and/or punctated decoration along with plain or dentate rocker stamping, with minor amounts of red filmed and brushed vessels also included (Ruby and Shriner 2005:556–557). From a collection of over 25,000 sherds, Ruby and Shriner (2005:556–557)

Table 3.1. Frequencies of Thin-Sectioned Vessels Analyzed from the Mann Site by Series, Type, and Temper

Series/Type	Local [Grog Temp]	Possibly Nonlocal *Tempers*				Totals
		LS	Grit	Chert	Sand	
Hopewellian Series						(18)
Hopewell Rims	3		1			4
Zoned Rocker Stamped	6					6
Plain Rocker Stamped	3					3
Red Filmed	5					5
Utilitarian Series						(16)
Cordmarked	9	1				10
Plain	2	3	1			6
Southeastern Series						(20)
Complicated Stamped	10				1	11
Coarse Simple Stamped	3	1				4
Fine Simple Stamped					5	5
Totals	41	5	1	1	6	54

Note: LS=Limestone

estimate that 91 percent derive from what they refer to as the Utilitarian Series and 6 percent from the Hopewellian Series. The remainder is assigned to the Southeastern Series, the vast majority of which (ca. 71 percent) are complicated stamped in sharp contrast to Ohio where this form of surface treatment is rare (Prufer 1968:14; Ruby and Shriner 2005:557–558).

In Table 3.1 the frequencies of pottery vessels from Mann whose thin sections were analyzed in this study are recorded by series, type, and temper group. These 54 thin sections were selected from among the 80 that were previously analyzed by Shriner (Ruby and Shriner 2005:559–560). Ruby and Shriner (2005:559) describe the 80 thin sections as "a nonrandom sample . . . reflecting the range rather than the proportional representation of variability present in the total assemblage." The distribution of the 80 thin sections by typological series is as follows: 24 Hopewellian, 36 Utilitarian, and 20 Southeastern (Ruby and Shriner 2005:560). As can be seen from Table 3.1, the present study includes 18 of the 24 Hopewellian, 16 of the 36 Utilitarian, and all 20 Southeastern vessels. The principal reason for excluding several of these thin sections pertained to their small sizes, which precluded their reliable analysis by the point counting procedures used in this study.

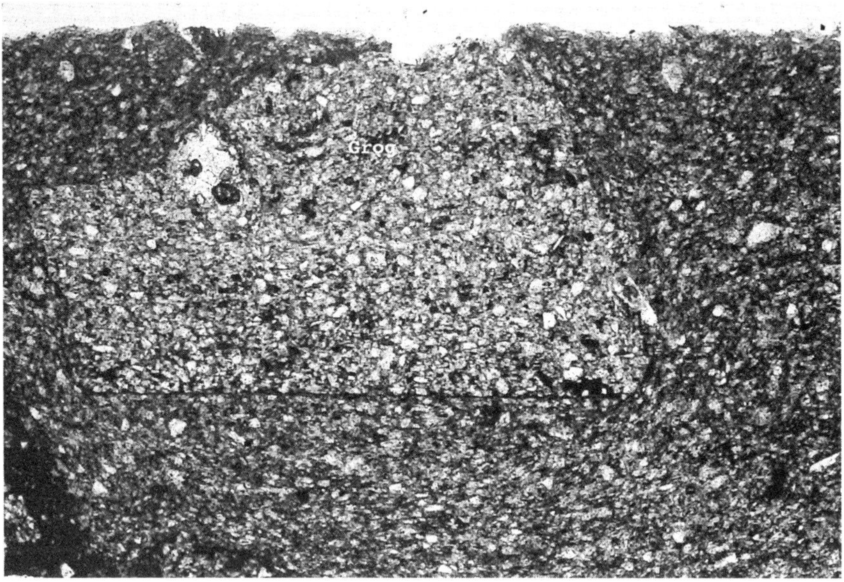

3.2. Photomicrograph of grog temper in local Vessel no. 1 (zoned rocker stamped) from the Mann Site taken at 10X magnification under plane-polarized light. Note lighter color and siltier texture of grog grain. Scale: grog= 1.78 mm.

Local Pottery at Mann

Initial, get-acquainted observations of several thin sections confirmed that, as reported by Kellar (1979:103), the preponderance of Mann pottery is characterized by "clay tempering," i.e., grog. It was on this basis that 41 of the 53 vessels were recorded as "Local" in Table 3.1, while the 13 vessels classed as "Possibly Nonlocal" all possess tempers other than grog, i.e., 2 have grit (1 syenitoid, 1 chert), 5 have limestone, and 6 are sand/fine-grit tempered. As in the case of the Ohio samples, these latter vessels were set aside for evaluation at the end of this chapter after the quantitative parameters of the locally made vessels had been established.

It was concerning the issue of grog recognition that the point count results of this study differ from those of Shriner's analysis. Because of difficulties in distinguishing natural argillaceous inclusions and iron stains from grog temper, Shriner elected not to attempt to distinguish the grog, instead recording it as matrix (Ruby and Shriner 2005:561). By contrast, despite the reality of this problem, but convinced that real grog was present in this sample (Figure 3.2), I attempted to distinguish grog whenever encountered using principles outlined in Chapter 1.

After point counting was completed, the compositional data for individual grog-tempered vessels were assembled by ceramic series as recorded in Tables 3.2 to 3.5, and the mean values for each of the series were calculated. At this stage, a perusal of the comparative series means revealed that, unlike the situation in Ohio, no significant differences in body values between the series existed. Indeed, if Cordmarked vessel no. 33 is disregarded as either intrusive or aberrant, the mean temper amounts in the Hopewellian and Utilitarian series are virtually identical, as are the temper size indices (cf. Tables 3.2 and 3.3). The mean body values for the Complicated Stamped vessels (Table 3.4) are also closely similar to those of the Hopewellian and Utilitarian series, but the Coarse Simple-Stamped vessels average somewhat higher temper percentages (Table 3.5). With only three such vessels in the sample, however, this difference can probably best be attributed to sampling error.

A similar comparison of the three series in regard to paste composition reveals even greater uniformity. Only two vessels stand out—Hopewell rim no. 18 and Cordmarked vessel no. 40 (see Tables 3.2 and 3.3)—both of which have unusually high percentages of sand (22 percent and 16 percent, respectively). It is possible that sand grains were added to complement the grog temper in these vessels or that a sandy clay was used in their manufacture. In either case local production rather than importation is the more parsimonious view in the absence of any evidence of an external source for them. As a consequence of these findings, the mean compositional values for the 41 grog-tempered vessels are accepted as a reasonably reliable compositional signature for pottery made locally at the Mann site.

By employing the grand mean of the 41 grog-tempered vessels as the basis for establishing a two-sigma standard (i.e., 95 percent confidence limits) for the site, it can be seen that the composition of the site's pottery is surprisingly uniform (the grand means are recorded in the bottom lines of Tables 3.2 to 3.5). Indeed, only 5 of the 41 vessels have compositional values falling beyond two-standard deviations of the grand mean. Two of these vessels—Hopewell rim no. 18 (see Table 3.2) and Cordmarked vessel no. 40 (see Table 3.3)—have unusually high sand values for both body and paste, indicating that they were made from sandier clays than the other local vessels or possibly had some sand added as temper. Whether these vessels should be regarded as simply atypical local products or imports is an open question. Complicated Stamped vessel no. 68 (see Table 3.4) has a slightly elevated sand-size index that can reasonably be attributed to sampling error, i.e., the chance occurrence of a large grain or two in a thin section with a low frequency (only 2 percent) of sand. The same logic applies to the anomalous temper size index for vessel no. 18 (see Table 3.2). The fourth vessel with an atypical compositional value—Cordmarked vessel no. 33 (see Table 3.3)—stands out in possessing a uniquely high incidence of temper (21 percent versus the grand mean of only 6.2 percent). It is possible that this vessel was made elsewhere, but its elevated temper incidence is here viewed as more likely attributable to idiosyncratic local behavior in light of the completely "local" paste values. The fifth vessel—Hopewell rim no. 23—was made from an unusually silty paste that will be shown to occur on the lower terrace below the site (see next section and Table 3.6). In sum, in the absence of an external site that can be cited as a source for any of the five vessels with anomalous compositional values, their local manufacture is accepted as most likely.

Accepting, then, the grand mean for the 41 grog-tempered vessels as providing a reasonably reliable representation of the composition of pottery made locally at the Mann site, two generalizations may be offered pertaining first to body, then to paste. The salient body property of Mann pottery is the existence of grog temper (see Figure 3.2). More than the mere presence of grog, however, Mann pottery composition is distinctive in being characterized by an unusually low incidence of temper, an average volume of only 6 percent (see Tables 3.2–3.5). In my experience in dealing with tempered pottery of any kind, it is rare to encounter vessels that do not possess double-digit temper volumes (e.g., cf. Tables 2.2–2.12). Among the 41 local Mann vessels, only 7 have more than 9 pecent temper and only one of those, Cordmarked vessel no. 33, has more than 13 percent grog (see Table 3.3).

With respect to paste, the compositional evidence indicates that, for the most part, Mann pottery was made from relatively fine-grained alluvial sediments, averaging ca. 88 percent matrix, 8 percent silt, and 4 percent sand (see Tables 3.2–3.5). As noted above, there are two notable exceptions among the 41 local vessels—Hopewell rim no. 18 and Cordmarked vessel no. 40—both characterized by unusually sandy pastes that could be attributed either to the use of distinctive, sandy clays or to the intentional addition of some sand temper. Whether these vessels were produced locally (the view currently accepted) or imported is presently an open question. It is also relevant to note that the paste values for Mann pottery fall fully within the range of variation documented for the Hopewellian sites of southern Ohio (cf. Table 2.10) and thus may be regarded as part of a broad regional compositional pattern that characterizes the central Ohio Valley region.

Local Sediments Used for Pottery Manufacture at Mann

Several of the thin sections provided by Dr. Shriner were derived from natural clayey sediments collected in and around the Mann site, which is located on

Table 3.2. Body and Paste Values by Vessel for the Hopewellian Series at the Mann Site

Values in bold fall outside the relevant ± two standard deviation ranges shown in the shaded cells. (Chert-tempered Hopewell Rim #19 excluded; see Table 3.7.)

		Body					Paste			
Type	Thin Section No.	% Matrix	% Sand	% Temper	Temper Size Index	Temper Type	% Matrix	% Silt	% Sand	Sand Size Index
Hopewell Series										
Zoned Rocker Stamped	1	87	1	12	3.87	Grog	92	7	1	2.00
Zoned Rocker Stamped	2	96	2	2	3.33	Grog	85	13	2	1.33
Zoned Rocker Stamped	3	96	2	2	3.25	Grog	90	8	2	1.00
Zoned Rocker Stamped	6	95	2	3	3.67	Grog	87	11	2	2.00
Zoned Rocker Stamped	7	92	5	3	4.00	Grog	90	5	5	1.17
Zoned Rocker Stamped	8	88	7	5	3.50	Grog	80	12	8	1.18
Plain Rocker Stamped	9	93	2	3/2	3.75/4.67	Grog/LS	88	10	2	1.00
Plain Rocker Stamped	10	98	1	1	3.00	Grog	92	7	1	1.00
Plain Rocker Stamped	11	93	2	5	4.00	Grog	91	7	2	1.67
Hopewell Rim	18	**75**	**21**	4	**5.00**	Grog	**71**	7	**22**	1.83
Hopewell Rim	20	92	1	7	3.38	Grog	91	8	1	1.00
Hopewell Rim	23	89	0	11	3.43	Grog	84	**16**	0	—
Red Filmed	12	89	2	9	4.17	Grog	91	7	2	2.00
Red Filmed	13	88	2	10	3.13	Grog	89	8	3	2.67
Red Filmed	14	94	1	5	3.43	Grog	92	7	1	1.00

Red Filmed	15	91	3	6	3.50	Grog	88	9	3	1.25
Red Filmed	16	95	0	5	3.33	Grog	96	4	0	—
Mean ±1 Std Deviation	[n=17]	91.2±5.3	3.2±4.9	5.6±3.2	3.63±.48		88.1±5.7	8.6±3.0	3.3±5.2	1.47±.52
Local Mean and Std Deviation	[n=41]	90.1±5.1	3.7±4.6	6.2±4.1	3.66±.40		87.6±5.1	8.5±2.8	3.9±4.7	1.65±.53
Two-Sigma Ranges	[n=41]	79.9-100	0-12.9	0-14.4	2.86-4.46		77.4-97.8	2.9-14.1	0-3.3	1.00-2.71

Note: Mecn temper size applies to grog only.

Table 3.3. Body and Paste Values for Cordmarked and Plain Vessels at the Mann Site

Values in bold fall outside the relevant ± two standard deviation ranges shown in the shaded cells.

		Body					*Paste*			
Type	**Thin Section No.**	**% Matrix**	**% Sand**	**% Temper**	**Temper Size Index**	**Temper Type**	**% Matrix**	**% Silt**	**% Sand**	**Sand Size Index**
Cordmarked	26	93	1	6	3.83	Grog	90	9	1	1.00
Cordmarked	27	85	8	7	3.86	Grog	80	14	6	2.33
Cordmarked	28	92	0	8	4.00	Grog	92	8	0	—
Cordmarked	29	91	1	8	3.83	Grog	88	11	1	1.00
Cordmarked	32	88	11	1	4.00	Grog	84	4	12	1.17
Cordmarked	33	**77**	2	**21**	3.62	Grog	90	8	2	2.00
Cordmarked	37	94	2	2/2	3.00/2.75	Grog/Grit	89	9	2	1.00
Cordmarked	39	83	12	5	3.80	Grog	82	5	13	1.62
Cordmarked	40	82	**16**	2	3.00	Grog	**76**	8	**16**	1.58
Plain	46	86	3	11	3.69	Grog	87	10	3	2.50
Plain	52	89	3	8	3.50	Grog	92	5	3	2.00
Mean ±1 Std Dev	[n=11]	87.3±5.3	5.3±5.4	7.4±5.4	3.65±.35		86.4±5.2	8.3±2.9	5.3±5.6	1.62±.57
Possibly Nonlocal										
Plain	42	**78**	2	**20**	**2.60**	Syenitoid	90	8	2	1.00
Local Mean ±1 Std Dev	[n=41]	90.1±5.1	3.7±4.6	6.2±4.1	3.66±.40		87.6±5.1	8.5±2.8	3.9±4.7	1.65±.53
Two-Sigma Ranges	[n=41]	79.9-100	0-12.9	0-14.4	2.86-4.46		77.4-97.8	2.9-14.1	0-13.3	1.00-2.71

Note: Std Dev=Standard Deviation; Mean temper size applies to grog only

Table 3.4. Body and Paste Values by Vessel for the Complicated Stamped Series from the Mann Site

Values in bold fall outside the ± two standard deviation ranges shown in the shaded cells.

		Body					*Paste*			
Type	**Thin Section No.**	**% Matrix**	**% Sand**	**% Temper**	**Temper Size Index**	**Temper Type**	**% Matrix**	**% Silt**	**% Sand**	**Sand Size Index**
Complicated Stamped	63	85	2	13	4.00	Grog	90	8	2	1.67
Complicated Stamped	64	86	11	3	4.00	Grog	81	7	12	1.70
Complicated Stamped	65	96	3	1	3.00	Grog	83	14	3	1.67
Complicated Stamped	66	91	7	2	3.00	Grog	85	8	7	1.25
Complicated Stamped	67	94	1	5	3.67	Grog	87	11	2	2.00
Complicated Stamped	68	95	2	3	4.00	Grog	89	9	2	**3.00**
Complicated Stamped	69	92	4	4	3.75	Grog	81	14	5	1.80
Complicated Stamped	70	92	1	7	4.00	Grog	93	6	1	2.00
Complicated Stamped	72	92	4	4	4.00	Grog	88	8	4	1.60
Complicated Stamped	73	95	1	4	3.62	Grog	92	7	1	2.00
Mean ±1 Std Deviation	[n=10]	91.8±3.7	3.6±3.2	4.6±3.4	3.70±.40		86.9±4.3	9.2±2.9	3.9±3.4	1.87±.46
Possibly Nonlocal										
Complicated Stamped	71					Sand	*78*★	***2***★	***20***★	*1.50*★
Local Mean ±1 Std Deviation	[n=41]	90.1±5.1	3.7±4.6	6.2±4.1	3.66±.40		87.6±5.1	8.5±2.8	3.9±4.7	1.65±.53
Two-Sigma Ranges	[n=41]	79.9-100	0-12.9	0-14.4	2.86-4.46		77.4-97.8	2.9-14.1	0-13.3	1.00-2.71

★*Values in italics denote bulk composition*

Table 3.5. Individual Body and Paste Values for Coarse Simple-Stamped Vessels from the Mann Site

All values fall within the relevant ± two standard deviation ranges shown at the bottom of the table.
The compositional values for limestone-tempered, coarse simple-stamped vessel no. 77 are recorded in Table 3.7.

		Body					*Paste*			
Type	**Thin Section No.**	**% Matrix**	**% Sand**	**% Temper**	**Temper Size Index**	**Temper Type**	**% Matrix**	**% Silt**	**% Sand**	**Sand Size Index**
Coarse Simple Stamped										
Coarse Simple Stamped	74	95	0	5	3.60	Grog	94	6	0	—
Coarse Simple Stamped	75	86	2	12	3.53	Grog	90	7	3	1.75
Coarse Simple Stamped	76	85	2	8/5	4.00/2.00	Grog/LS	92	6	2	2.33
Mean ±1 Std Deviation	**[n=3]**	88.7±5.5	1.3±1.2	10.0±4.4	3.42±.26		92.0±2.0	6.3±0.6	1.7±1.5	2.04±.41
Local Mean ±1 Std Deviation	[n=41]	90.1±5.1	3.7±4.6	6.2±4.1	3.66±.40		87.6±5.1	8.5±2.8	3.9±4.7	1.65±.53
Two-Sigma Ranges	[n=41]	79.9–100	0–12.9	0–14.4	2.86–4.46		77.4–97.8	2.9–14.1	0–13.3	1.00–2.71

Note: Mean temper size applies to grog only.

Table 3.6. Bulk Composition of Natural Sediments from the Mann Site Vicinity

	Thin Section No.	% Matrix	% Silt	% Sand	Sand Size Index
Location 3					
Lower Terrace	3/3/4	85	14	1	1.50
Lower Terrace	3/4/5	86	13	1	1.33
Lower Terrace	3/5/6	81	17	2	1.00
Mean ±1 Std Deviation	**[n=3]**	84.0±2.6	14.7±2.1	1.3±0.6	1.28±.25
Location 5					
High Terrace	5M UR 1/2	78	20	2	2.75
High Terrace	5M UR 3/4	78	20	2	2.00
High Terrace	5M UR 5	76	23	1	1.00
Mean ±1 Std Deviation	**[n=3]**	77.3±1.2	21.0±1.7	1.7±0.6	1.92±0.9
"Mann Site Soil"	Station 9	90	9	1	1.00
Mann Paste Mean	**[n=41]**	88	8	4	1.66

the higher of two glacial outwash terraces (see Ruby and Shriner 2005:555 and 559). Seven of these thin sections were subjected to the same point counting procedure as the pottery thin sections. According to information provided by Dr. Shriner (e-mail of June 19, 2003), the sediments analyzed were derived from two loci as follows: Location 3="clay-rich sediments from the lower terrace area," and Location 5="clay deposits from a creek bed local to the Mann site" (see map in Ruby and Shriner 2005:555). The Location-3 sediments were collected high within the lower terrace near the boundary with the upper terrace (Ruby and Shriner 2005:559), while the Location-5 sediments were collected from a depth of four meters within a creek bed that had cut deeply into fluviolacustrine sediments that blanket the upper terrace where the Mann site resides (Ruby and Shriner 2005:559). In addition to three thin sections from each of the above two loci, a seventh thin section bearing the label "Mann Site Soil" was also analyzed. The results of the analysis of these seven sediment samples are presented in Table 3.6.

From the data in Table 3.6, it appears that the sediment represented in a thin section labeled "Mann Site Soil" most closely matches the mean paste value for local Mann pottery. This sediment was presumably that derived from Location 9, a one-meter deep posthole on the Mann site itself (Ruby and Shriner 2005:559). The sediments from Location 3, while siltier than the mean for Mann pottery, clearly have a range of variation that overlaps the latter, thus cannot be ruled out as a clay source for Mann potters. Indeed, Hopewell rim no. 23, with its distinctively high silt content appears to have been made from exactly such clays (see Table 3.2). Location 5, by contrast, yielded clays that appear to be too silty to be considered a source for Mann potters. Since the stream on the high terrace from which the Location-5 sediments were collected has its headwaters in the adjacent loess-covered uplands (see Ruby and Shriner 2005:555), it is likely that its valley was filled with redeposited loess, a high-silt/low-sand sediment similar to that of the Location 5 samples. The logical conclusion from these data, that Mann potters were utilizing easily accessible, subsurface clays on and about the site itself, is fully consistent with the views expressed by Ruby and Shriner (2005:561).

Evaluating the "Possibly Nonlocal" Vessels

Thirteen of the 54 thin-sectioned vessels from the Mann site analyzed for this study were classed as "Possibly Nonlocal" because they were tempered differently from the grog-tempered local vessels (see Table 3.1). The tempers represented in these vessels, and their frequencies, are as follows: syenitoid (n=1); chert (n=1); limestone (n=5, one also with a unique calcareous sandstone); and sand/fine-grit (n=6). Consideration of the six sand/fine grit-tempered vessels (five have fine simple-stamped surface treatment and one is complicated stamped) will be deferred until Chapter 10 after discussion of possible Southeastern sources has been presented. Since the syenitoid-tempered, plain vessel no. 42 was previously assessed as a probable import from Ohio (see Table 2.22 and Table 3.3), the five limestone-tempered vessels and the chert-tempered Hopewell rim no. 19 are the remaining "Possibly Nonlocal" vessels to be considered here. The paste values for these six vessels are recorded in Table 3.7.

Employing the same approach used previously to evaluate the local versus nonlocal status of such vessels in Ohio, the compositional parameters considered most important and relevant to this task are those of paste. As can be seen from Table 3.7 all paste values for the six vessels except one—the sand size index for coarse simple-stamped vessel no. 77—fall fully within the two-sigma paste ranges for the 41 local Mann vessels. In the latter case a single coarse sand grain in a small thin section produced this elevated size index. Almost certainly this value can be attributed to sampling error. Thus, the current compositional evidence provides no basis for rejecting the view that the

Table 3.7. Paste Values for Five Limestone-Tempered Vessels and the Chert-Tempered Hopewell Rim from the Mann Site Compared with the Two-Sigma Paste Ranges for the Local Mann Pottery

Value in bold exceeds the ± two standard deviation range shown in the shaded cell.

Pottery Type	Thin Section No.	% Matrix	% Silt	% Sand	Sand Size Index
Limestone Tempered					
Cordmarked	25	93	5	2	1.00
Plain	49	82	9	9	1.33
Plain	50	95	5	0	—
Plain	54	97	3	0	—
Coarse Simple Stamped	77	94	4	2	**3.00**
Mean ±1 Std Deviation	[n=5]	92.2±5.9	5.2±2.3	2.6±3.7	1.78±1.1
Two-Sigma Ranges	[n=5]	80.7–100	0.6–9.9	0–10	1.00–3.98
Hopewell Rim, Chert Tempered	19	92	6	2	2.00
Mann Mean ±1 Std Deviation	N=41	87.6±5.1	8.5±2.8	3.9±4.7	1.65±.53
Mann Two-Sigma Ranges	N=41	77.4–97.8	2.9–14.1	0–13.3	1.00–2.71

five limestone-tempered vessels and the unique chert-tempered vessel were manufactured locally at Mann.

Mann Pottery Recovered Elsewhere in the Eastern Woodlands

In the course of this study, seven sherds of grog-tempered pottery were encountered at five Middle Woodland sites outside of Indiana (Table 3.8). At each of the five sites these particular sherds stood out because of their presence in ceramic assemblages that were otherwise grit or sand tempered. Since, on stylistic grounds, they all *could* be derived from the Mann site, the critical test of their intrusive status revolves around the compositional evidence, which is now addressed.

The seven grog-tempered sherds involved derive from the following sites: Hopewell and Liberty in Ohio; Garden Creek Mound No. 2 in North Carolina; Leake in north Georgia; and Pinson Mounds in west Tennessee (see Table 3.8). In comparing the compositions of these vessels to those of local Mann

Table 3.8. Body and Paste Values for Seven Grog-Tempered Vessels Recognized as Intrusive to Their Respective Sites Compared with the Two-Sigma Body and Paste Ranges for Local Mann Pottery

Value in bold font falls outside the relevant ± two standard deviation range for Mann shown in the shaded cell.

		Body					Paste			
Site & Pottery Type	**Thin Section No.**	**% Matrix**	**% Sand**	**% Temper**	**Temper Size Index**	**Temper Type**	**% Matrix**	**% Silt**	**% Sand**	**Sand Size Index**
Hopewell, OH (Plain)	33-154	96	1	3	3.29	Grog	85	14	1	1.33
Liberty, OH (Cordmarked)	33-149	80	1	**19**	4.40	Grog	88	11	1	1.00
Liberty, OH (Cordmarked)	33-237	88	7	5	3.86	Grog	83	10	7	2.10
Garden Creek, Mound 2 (Burnished Pl.)	31-26	94	1	5	3.08	Grog	92	7	1	2.00
Leake, GA (Rocker Stamped)	9-200	86	6	8	4.17	Grog	90	3	7	2.33
Leake, GA (Complicated St.)	9-273	87	1	12	4.40	Grog	88	11	1	1.00
Pinson Mounds, TN (Red slipped)	40-66	83	7	10	4.00	Grog	80	12	8	1.10
Mean ±1 Std Deviation	**n=7**	87.7±5.7	3.4±3.0	8.9±5.5	3.89±.52		86.6±4.2	9.7±3.6	3.7±3.4	1.55±.57
Mann Mean ±1 Std Deviation	**n=41**	90.1±5.1	3.7±4.6	6.2±4.1	3.66±.40		87.6±5.1	8.5±2.8	3.9±4.7	1.65±.53
Mann Two-Sigma Ranges	**n=41**	79.9-100	0-12.9	0-14.4	2.86-4.46		77.4-97.8	2.9-14.1	0-13.3	1.00-2.71

Note: St=Stamped

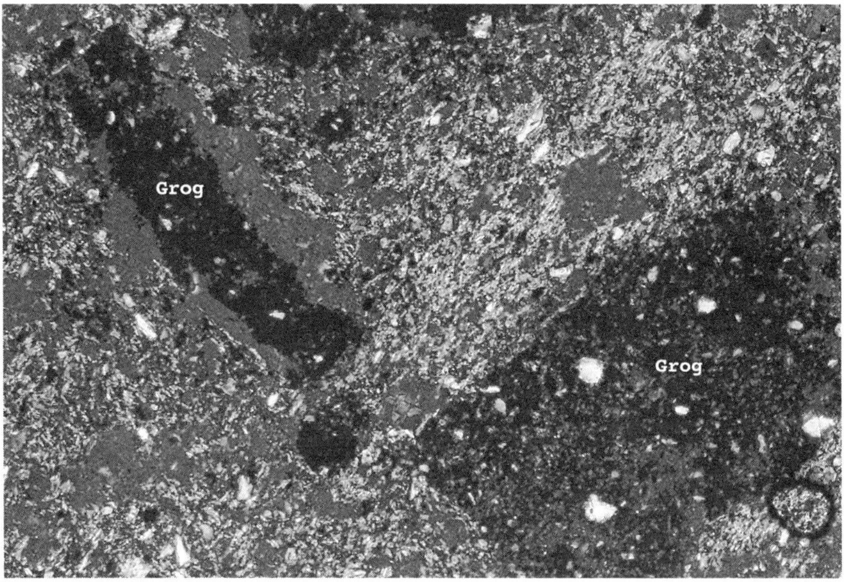

3.3. Photomicrograph of grog temper in Vessel 31–26 from Garden Creek taken at 25X magnification under crossed polars. Scale: largest of grog (upper left)= .45 mm. This vessel is considered an import from the Mann site.

pottery, it should be noted that *both* body and paste values are relevant because all vessels share the common property of grog temper. As can be seen from Table 3.8, all compositional values for these seven vessels, with but the single exception of "percent Temper" in the body of vessel 33-149 from Liberty, fall fully within the two-sigma/95 percent confidence limits of Mann ceramics. That the unusually high temper percent of this vessel should disqualify it as a candidate for Mann derivation is doubtful in light of evidence that even higher temper percentages may sometimes occur at Mann (vessel 33 has 21 percent grog, Table 3.3). Because of the unusually low incidence of grog that generally characterizes Mann pottery, it can reasonably be suggested that such elevated temper counts reflect sampling error, i.e., an unusual density of grog in a part of the vessel that does not fairly represent the composition of the whole vessel.

Three of these vessels are deserving of special comment. The Garden Creek vessel (no. 31-26) is described by Keel (1976:118) as "a highly burnished necked jar fragment (Plate 18c) which was not typical of the eastern Tennessee area," thus was tentatively assigned to a Florida type, Crystal River Plain. There is no doubt that this vessel is grog tempered (Figure 3.3), a practice virtually unheard of in North Carolina and peninsular Florida (relatively rare occurrences of grog tempering do occur in Swift Creek contexts

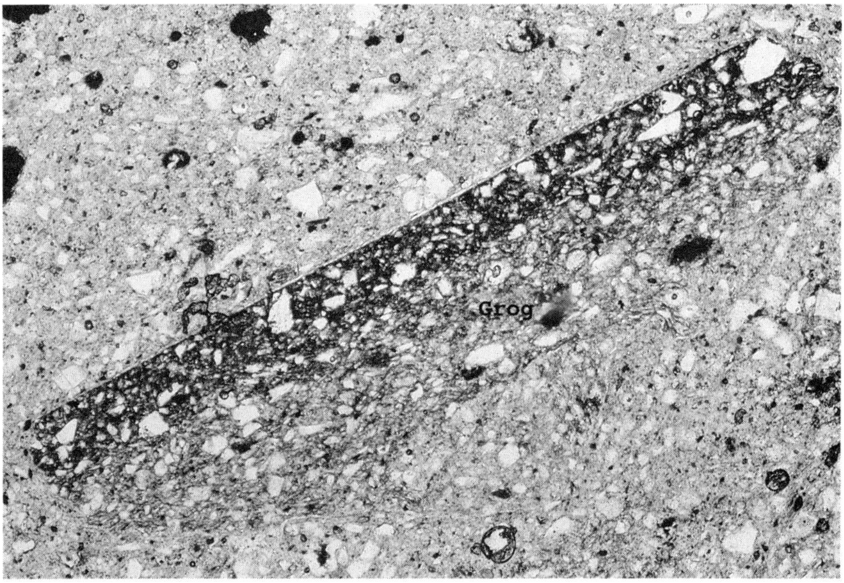

3.4. Photomicrograph of grog temper in red-filmed Vessel 40-66 from Pinson Mounds taken at 10X magnification under plane polars. Scale: grog= 1.90 mm. This vessel is considered an import from the Mann site.

in the Florida panhandle; Cordell, personal communication). Since all of its compositional parameters of both paste and body fall fully within the two-sigma ranges of Mann ceramics (see Table 3.8), this vessel seems an excellent, if surprising, candidate for an import from the Mann site.

The second of the seven vessels deserving of special comment is complicated stamped vessel 9-273 from the Leake site. From a substantial sample of 47 Swift Creek Complicated Stamped vessels from this site that were thin sectioned, this is the only one that was grog tempered (Stoltman 2010). Considering that all of its compositional parameters of both paste and body fall within the two-sigma limits of Mann pottery, it must be viewed as an excellent candidate for a Mann manufacture. This is especially noteworthy, for it accentuates the point that the Mann site was a locus of manufacture of complicated stamped pottery well to the north of the Swift Creek heartland.

The third grog-tempered vessel deserving of special comment is the red-filmed vessel from Pinson Mounds (Figure 3.4). In an earlier publication (Stoltman and Mainfort 2002), this vessel was identified as exotic to Pinson Mounds and suspected on stylistic grounds of being derived from the Mississippi Valley. Thus, it was referred to as "Larto Red Filmed," a type characteristic of that region (cf. Phillips, Ford, and Griffin 1951:102–105). Subsequently, the Mann thin sections were analyzed for this report, as were thin sections from nine Marksville Incised vessels from three Yazoo Basin sites to the west, which were graciously provided by John Connaway. Table 3.9 records the body and paste values for each of the nine vessels by site. The two-sigma ranges for body and paste values for the nine Marksville Incised vessels are also recorded near the bottom of the table for comparison with the values for vessel 40-66 from Pinson Mounds. From Table 3.9, it can be readily seen that most of the compositional values of vessel 40-66 fall beyond the 95 percent confidence limits of the sample of classic Middle Woodland Yazoo Basin sherds.

Accepting for the moment that the current sample of thin-sectioned vessels from the Yazoo Basin fairly represents the composition of Middle Woodland pottery from that region, it is difficult to see the Pinson Mounds vessel as derived from the Yazoo Basin. The attribution of the Mann site as the probable source for this vessel represents a dramatic reassessment of its nonlocal status and, of course, forcefully contradicts the INAA data that once identified this vessel as a local product of Pinson Mounds (Mainfort et al. 1997).

Based upon these findings, the Mann site appears to have been one of the most active participants in the Hopewell Interaction Sphere, at least insofar as this can be documented through the circulation of pottery vessels. On the one hand the unparalleled abundance of complicated stamped pottery at Mann—over 600 sherds (Kellar 1979:103)—suggests that it was the recipient of recurrent and/or extensive visitation by people from the Swift Creek heartland (Williams and Elliot 1998). What is especially interesting about these complicated stamped vessels at Mann is that, with but one exception, all appear to have been made locally. On this point, Ruby and Shriner (2005:569) and the author are in complete accord. This conclusion is based upon petrographic observations on thin sections from 11 complicated stamped sherds, 10 of which are grog tempered, thus accepted as local manufactures (see Table 3.4).

Rejection of direct importation of complicated stamped pottery from the Southeast to account for the occurrence of most such vessels at Mann immediately raises a new issue, the origin of the carved paddles used in their manufacture. Ruby and Shriner (2005:571) consider this issue in terms of three alternatives: 1) local manufacture of paddles at Mann by people who had learned or procured access to the craft from Swift Creek people; 2) importation of the paddles from the Swift Creek province; and 3) manufacture of the pottery at Mann by Swift Creek visitors. All are certainly viable possibilities, especially the latter two, which are consistent with evidence from

Table 3.9. Body and Paste Values for Marksville Incised Vessels from the Yazoo Basin of Mississippi Compared with Red-Slipped Vessel 40-66 from Pinson Mounds

Pinson Mounds Values in bold font fall outside the two-sigma/95% confidence limits for the Yazoo Basin pottery.

		Body					*Paste*			
Site/Type	**Thin Section No.**	**% Matrix**	**% Sand**	**% Temper**	**Temper Size Index**	**Temper Type**	**% Matrix**	**% Silt**	**% Sand**	**Sand Size Index**
Dickerson (22Co502)										
Marksville Incised	22-21	94	0	6	3.17	Grog	96	4	0	—
Marksville Incised	22-22	91	0	9	3.40	Grog	98	2	0	—
Marksville Incised	22-23	91	1	8	3.30	Grog	96	3	1	1.00
Mean ±1 Std Deviation	[n=3]	92.0±1.7	0.3±0.6	7.7±1.5	3.29±.12		96.7±1.2	3.0±1.0	0.3±0.6	1.00±0
Dry Bayou (22Co266)										
Marksville Incised	22-27	87	0	13	3.75	Grog	98	2	0	—
Marksville Incised	22-28	94	0	6	3.12	Grog	96	4	0	—
Marksville Incised	22-29	91	0	9	3.73	Grog	98	2	0	—
Mean ±1 Std Deviation	[n=3]	90.7±3.5	0±0	9.3±3.5	3.53±.36		97.3±1.2	2.7±1.2	0±0	—
Rochdale (22Go654)										
Marksville Incised	22-24	90	0	10	3.88	Grog	89	11	0	—
Marksville Incised	22-25	85	0	15	2.55	Grog	98	2	0	—
Marksville Incised	22-26	89	2	9	3.72	Grog	88	9	3	1.00

Mean ±1 Std Deviation	[n=3]	88.0±2.6	0.7±1.2	11.3±3.2	3.38±.73	91.7±5.5	7.3±4.7	1.0±1.7	1.00±0
Grand Mean ±1 Std Deviation	[n=9]	90.2±3.0	0.3±0.7	9.5±3.0	3.40±.42	95.2±3.9	4.3±3.4	0.5±1.0	1.00±0
Yazoo Two-Sigma Ranges	[n=9]	84.2-96.2	0-1.7	3.5-15.5	2.56-4.24	87.4-100	0-11.1	0-2.5	1.00
Pinson Mounds	40-66	**83**	**7**	10	4.00	**80**	**12**	**8**	**1.10**

the Swift Creek heartland that paddles circulated between sites (Stoltman and Snow 1998:147–150).

Besides being uniquely rich in complicated stamped pottery, the Mann site stands out among sites in this study as the second-most prolific exporter of pottery vessels to other Middle Woodland sites (see the discussion of the Leake site in Chapter 10). If the evidence presented in Table 3.8 can be accepted at face value, vessels manufactured at Mann reached sites as far apart as western Tennessee and the Appalachian highlands of North Carolina, a truly unique state of affairs.

4
Southern Illinois

Two sites near the lower Ohio River in southern Illinois are included in the present study, the Rutherford Mound in Hardin County and 11Mx109 in Massac County (see Figure 3.1). Each site will be considered in turn.

The Rutherford Mound

Melvin L. Fowler excavated the Rutherford Mound in 1954 in behalf of the Illinois State Museum (Fowler 1957). From a prepared floor at the base of the mound and from the mound fill, 35 burials were encountered. Several classic Hopewellian artifact types were recovered, including ear spools, panpipes, a celt, and a disc of copper; cut bear canine teeth; pearl, shell, and copper beads; as well as a wide range of other mortuary items and exotic materials.

Petrography was employed to determine the composition of only two of the pottery vessels from the site, so this cannot be considered a comprehensive report of the site. The two vessels that were thin sectioned include the well known, nearly complete, simple-stamped jar with a tetrapodal base that was recovered near the left shoulder of Burial 6 (Fowler 1957:27) and a nearly complete cordmarked jar that was recovered from the mound fill (Fowler 1957:10–11; Plate Ia). The reason for the inclusion of these two vessels revolves around suspicions of their intrusive status: the simple-stamped vessel with a tetrapodal base is stylistically anomalous for southern Illinois, while Prufer's suggestion (1965:22) that the latter vessel may be the westernmost occurrence of the type McGraw Cordmarked raises the interesting possibility that this vessel was imported from Ohio.

Observations on the simple-stamped vessel (thin section no. 11-242) confirm that it is, indeed, sand tempered as described by Fowler (1957:27). That is to say, it is characterized by relatively abundant fine to medium inclusions (sand size index=1.49) that are preponderantly monocrystalline quartz. Considering its stylistic properties, a Southeastern derivation for this vessel is deemed highly probable. We will return to evaluate this vessel further after potential Southeastern sources have been considered in Chapter 10.

The cordmarked vessel (Plate Ia in Fowler 1957:11; thin section no. 11-243 in this study), by contrast, is characterized by limestone temper that has been leached out. While citing affinities to both Weaver ceramics of the Illinois Valley and local Lewis-Raymond types, Fowler (1957:12) ultimately opined that it was of late Middle Woodland affiliation. Although uncommon in the lower Ohio Valley, limestone temper is present in Middle Woodland contexts within the local Crab Orchard tradition (e.g., Butler and Jefferies 1986; Butler and Scott 2008). Indeed, two limestone-tempered sherds were surface collected from the nearby Kaegi site, which Fowler (1957:41) attributed "to the same relative time period and culture" as the Rutherford mound.

The physical composition of vessel 11-243 as determined by petrographic observation is as follows:

Paste: 90 percent matrix; 9 percent silt; 1 percent sand; sand size index=1.00
Body: 96 percent matrix; 1 percent sand; 3 percent limestone temper; temper size index=3.75

To evaluate effectively the "Prufer hypothesis" of the Ohio origin of this vessel, a two-way compositional comparison would be required: (1) with local, limestone-tempered vessels, and (2) with limestone-tempered vessels from Ohio that had "local" compositional values there. Because no local vessels from Rutherford were available for this study, it is currently impossible to attempt the first of these comparisons. As for the second, that can be attempted by comparing the paste values for this vessel with those of the 18 limestone-tempered vessels from Hopewell sites in southern Ohio and the five Mann vessels that were previously accepted, if tentatively, as local to those regions. By referring to Tables 2.20, 3.7, and 8.4, it can be seen that vessel 11-243 has paste values that fall fully within the two-sigma limits of the paste values for Mann and the Ohio sites. Thus, there appears to be a general regional pattern of paste composition for limestone-tempered vessels of the Ohio Valley that overlap so much that specific source determination may be problematic. We shall return to this issue in Chapter 8 after limestone-tempered pottery in the Southeast is considered and body values can be added to the discussion.

11Mx109

Excavated by Jon Muller in 1974 as one of Southern Illinois University's annual field schools, the artifacts recovered from 11Mx109 languished in storage until 2004 when an SIU graduate student, Mackenzie Caldwell (later Rohm) commenced an analysis of the site's pottery for her Master's thesis. With the encouragement of Brian Butler, Rohm's major professor, who brought these materials to at least two professional meetings seeking opinions on their cultural affiliation, the uniqueness of the site became apparent. As described by Rohm (2008:3): "This assemblage does not fit into the local ceramic sequence and provides a striking example of discontinuity in the traditional understanding of the culture history of southern Illinois and the lower Ohio River valley."

The inclusion of 11Mx109 in this study is based upon petrographic observations of twenty sherds that were analyzed to assist Rohm in the preparation of her Master's thesis. 11Mx109 is a multicomponent habitation site, with the thin-sectioned sherds representing ten vessels each from the Middle Woodland and Late Woodland components. Consonant with the focus of this study, only the Middle Woodland pottery is described. These data are reported here in order that they may be readily viewed in the larger context provided by this study. In addition their interpretation now has the benefit of a much larger comparative database than was available at the time Rohm (2008) was writing her thesis.

The compositional values for the ten Middle Woodland vessels that were thin sectioned are recorded in Table 4.1.

The uniformity of these vessels, as reflected in their body and paste values is noteworthy. Indeed, all compositional values except two grain-size indices—the temper-size index for vessel 11-338 and the sand-size index for vessel 11-333—fall within the 95 percent confidence limits for the site (see Table 4.1). In both instances the elevated grain-size indices apply to vessels that have the lowest frequencies in the sample for temper (5 percent grog for vessel no. 11-338) or sand (1 percent sand for vessel no. 11-333), making them especially susceptible to sampling error when a single large grain or two appeared in the count. The relative uniformity of the 11Mx109 ceramic sample, small though it may be, favors accepting the values presented in Table 4.1 as providing a reliable compositional signature for the site's Middle Woodland pottery.

The obvious and most relevant comparative data for attempting to place the 11Mx109 ceramic assemblage in cultural context derive from the Mann site, which is located less than 80 km (roughly 70 miles) upstream along the lower Ohio River (See Figure 3.1). To facilitate comparing the ceramic compositions of the two sites, the means, standard deviations, and two-sigma ranges for the Mann pottery are recorded at the bottom of Table 4.1. The most obvious similarity between the ceramics of the two sites, of course, is that grog-tempered bodies characterize both. While sharing the use of grog

Table 4.1. Body and Paste Values by Vessel for 11Mx109 Pottery Compared with Mann Means

Values in bold font fall outside the relevant ± two standard deviation ranges shown in the shaded cells.

		Body					*Paste*			
Type	Thin Section No.	% Matrix	% Sand	% Temper	Temper Size Index	Temper Type	% Matrix	% Silt	% Sand	Sand Size Index
Plain Rim	11-330	89	1	10	3.20	Grog	86	13	1	1.00
Plain	11-331	81	1	18	3.34	Grog	81	18	1	1.50
Plain	11-332	84	1	15	3.05	Grog	82	17	1	1.00
Plain base	11-333	77	1	22	3.14	Grog	90	9	1	**3.00**
Zoned Rocker Stamped	11-334	90	1	9	3.58	Grog	92	7	1	1.00
Plain Neck	11-335	82	1	17	3.59	Grog	93	6	1	1.00
Zoned Dentate	11-336	83	4	13	3.21	Grog	94	1	5	1.50
Plain	11-337	78	0	22	3.17	Grog	87	13	0	—
Cordmarked	11-338	81	14	5	**4.00**	Grog	78	7	15	1.38
Plain Rim	11-339	76	14	10	3.47	Grog	80	4	16	1.38
Mean ±1 Std Dev	**[n=10]**	82.1±4.7	3.8±5.5	14.1±5.7	3.38±.29		86.3±5.8	9.5±5.6	4.2±6.1	1.42±.63
Two-Sigma Ranges	**[n=10]**	72.7-91.5	0-14.8	2.7–25.5	2.80-3.96		74.7-97.9	0-20.7	0-16.4	1.00-.68
Mann Mean ±1 Std Dev	**[n=41]**	90.1±5.1	3.7±4.6	6.2±4.1	3.66±.40		87.6±5.1	8.5±2.8	3.9±4.7	1.65±.53
MannTwo-Sigma Ranges	**[n=41]**	79.9-100	0-12.9	0-14.4	2.86-4.46		77.4-97.8	2.9-14.1	0-13.3	1.00-2.71

Note: St=Standard; Dev= Deviation

temper, it is interesting to note that the amount of temper differs significantly between the sites: for Mann the mean percent temper is 6.2 ± 4.1 versus 14.1 ± 5.7 for 11Mx109 (see Table 4.1). The application of a *t*-test of the significance of the difference between these means results in a value of $t=5.09$, which at 49 degrees of freedom has a probability of $<.001$, i.e., is highly significant. These data demonstrate that potters at the two sites used similar materials but different recipes in manufacturing their pottery containers.

Similarly, though less dramatically, the paste values differ between the two sites in that those of 11Mx109 are more variable than those of the Mann site—note the larger standard deviations associated with all of the paste indices for 11Mx109 (Table 4.1). This is reflected in the fact that at least five silt or sand values for 11Mx109 vessels fall outside the two-sigma limits for Mann, yet fall within the two-sigma limits for 11Mx109. Presumably these paste differences reflect the character of the local alluvial sediments at the sites. Based upon this compositional analysis, the ceramic industries at the two sites have been revealed to be closely similar technologically but somewhat different in terms of materials used. On the one hand, presumably because of compositional differences of local sediments, the pastes at the two sites differ somewhat, while on the other hand, the use of greater amounts of temper by the 11Mx109 potters presumably reflects a regionally distinct cultural practice. In conclusion, it is tempting to speculate that, considering the similarities between the ceramic industries of the two sites and the lack of identifiable antecedents for 11Mx109 within its region, the ultimate origin of the occupants of 11Mx109 was the Mann site.

5

The Blue Ridge Province of North Carolina and Tennessee

Astride the border between Tennessee and North Carolina, and extending a bit into north Georgia, lies the southern end of the Blue Ridge physiographic province (e.g., Fenneman 1938:163–194). The southern Blue Ridge is a rugged mountainous belt (more than 300 peaks rise above 5,000 feet) up to 113 km in breadth that is underlain by igneous and metamorphic crystalline rocks. This region is home to two of the more celebrated Middle Woodland sites in the Southeast, Garden Creek in North Carolina (Keel 1976) and Icehouse Bottom in Tennessee (Chapman 1973). These sites enjoy well-deserved celebrity because of the recovery, in carefully excavated contexts, of direct evidence of cultural contact with the Hopewellian centers of southern Ohio. In both cases this evidence consists of prismatic blades of Flint Ridge chalcedony (Keel 1976:136–142; Chapman 1973:92–95), a highly distinctive material that derives from the extensive quarries in Licking County, Ohio (Mills 1921). More germane to this study, however, is the presence at both sites of substantial numbers of sherds that bear stylistic features, such as rocker stamping and crosshatched rims, reminiscent of Ohio Hopewell pottery. The remainder of this chapter is devoted to the compositional analysis of a sample of these stylistically distinctive sherds in an effort to determine whether their suspected exotic origins can be confirmed.

Garden Creek Mound 2

The Garden Creek site is located on the Pigeon River about 2 miles west of the town of Canton in Haywood County in southwestern North Carolina (see Figure 3.1). The site consists of three mounds and two village areas within a 15-acre area along the south bank of the Pigeon River (Keel 1976: 65–67). In 1915 George G. Heye (1919) excavated one of the mounds, Mound 3. Subsequently, from 1965 to 1967, Bennie Keel, in behalf of the Research Laboratories of Anthropology of the University of North Carolina, con-

Table 5.1. Frequencies of Thin-Sectioned Vessels by Ceramic Series from Garden Creek Mound 2

Ceramic Series/Types	**Frequencies**
Connestee Series	[15]
Simple Stamped	12
Cordmarked	2
Plain	1
Hopewell Series	[10]
Rocker Stamped	7
Cross-Hatched Rim	1
"Turner Simple Stamped"	2
"Crystal River Plain"	1
Totals	26

ducted more intensive investigations focused primarily on Mound 2 (Keel 1976). The contents of Mound 2 are the subject of this report.

Petrographic analyses were conducted on thin sections prepared from 26 sherds, each derived from a different vessel (Table 5.1) from Garden Creek Mound 2. These sherds, all from the collections housed at the Research Laboratories of Anthropology at the University of North Carolina, were obtained through the good offices of the director of those laboratories, Vincas Steponaitis.

As recorded in Table 5.1, 15 of the 26 sherds are local types based upon stylistic considerations. An additional 10 sherds were derived from a group of nearly 100 identified as Ohio Hopewell types (Keel, 1976:118–120), while a unique "Crystal River Plain" vessel (Keel 1976:Plate 18c) was also included in the thin-section sample.

The first step in the petrographic analysis was a qualitative review of each thin section to gain a general impression of their mineralogical content. Because of the relatively fine size and mineralogical diversity of the inclusions, it was thought that the discrimination of temper—i.e., intentional human additives—from natural sand grains could not be accomplished objectively. Therefore, in the next step, rather than employing paste and body indices, the thin section contents were recorded as bulk compositions along with a mineralogical index (see Chapter 1 for a discussion of these indices).

In order to evaluate the possible exotic status of the Hopewell-series vessels, the next step involved the determination of the bulk and mineralogical composition of pottery that could be accepted without question as locally made to serve as a "local" standard. The 15 Connestee-series vessels provided the basis for this step.

The general mineralogical composition of these vessels was reviewed to see if any were distinctive outliers. Three vessels stood out in departing from the prevailing compositional pattern that characterized the other 12. Two types of inclusions immediately strike the eye when viewing these 12 thin sections: (1) relatively large grains composed mainly of strained, polycrystalline quartz sometimes accompanied by minor feldspar inclusions, and (2) smaller, scattered, monocrystalline grains of hornblende, strongly green in color and pleochroic in character (i.e., marked by color change upon rotation of the stage when viewed in plane polarized light). Additional mineral grains include feldspars in modest amounts (untwinned feldspar, probably orthoclase, lesser amounts of plagioclase, and only rare microcline) and small amounts of both muscovite and biotite micas. It is likely that the preponderance of these grains were derived from two or three different metamorphic rocks, metagranite, metaquartzite, and amphibolite (Figure 5.1).

Temporarily holding aside the three anomalous Connestee vessels as potentially nonlocal, the remaining 12 Connestee vessels appear to be compositionally homogeneous (only two have a compositional value that exceeds the group two-sigma range, both by less than one percentage point; Table 5.2), thus are accepted as the best candidates from which to derive a compositional signature for locally made pottery at Garden Creek.

The bulk and mineralogical compositions of these 12 vessels, as well as for the three anomalous Connestee vessels, are recorded in Table 5.2. Considering bulk composition first, two properties stand out as distinctive of the 12 Connestee vessels, (1) the high incidence of sand-size grains (a mean of 34.4 percent$\pm$ 5.2) and (2) the relatively modest sizes of these mineral inclusions (the mean size of the sand grains is 1.93$\pm$.16), which is much smaller than occurs in Hopewellian pottery in Ohio (cf. Table 2.13). Mineralogically, as described above, polycrystalline quartz and hornblende stand out prominently in these thin sections (e.g., Figure 5.1). This is expressed clearly in Table 5.2 by the high incidence of both polycrystalline and FMM grains and the relative paucity of monocrystralline quartz.

With this local signature thus established, let us turn to an evaluation of the potentially exotic status of the ten Hopewell-series vessels (Figure 5.2). Starting again with a general review of the mineralogical composition of these vessels, three at once stand out as distinctive. These will be considered after discussion of the remaining eight Hopewell-series vessels has been completed. From Table 5.3, it can be seen that the compositions of these eight vessels are closely similar to those of the 12 local Connestee vessels (see Table

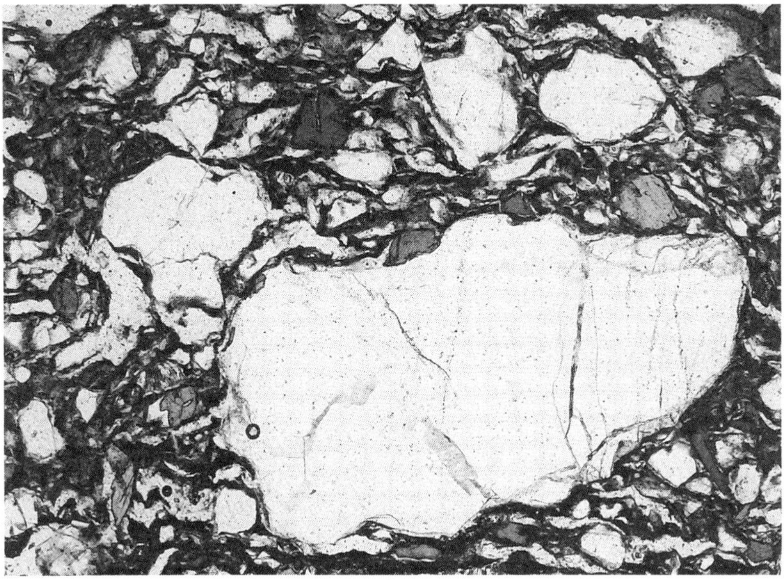

A

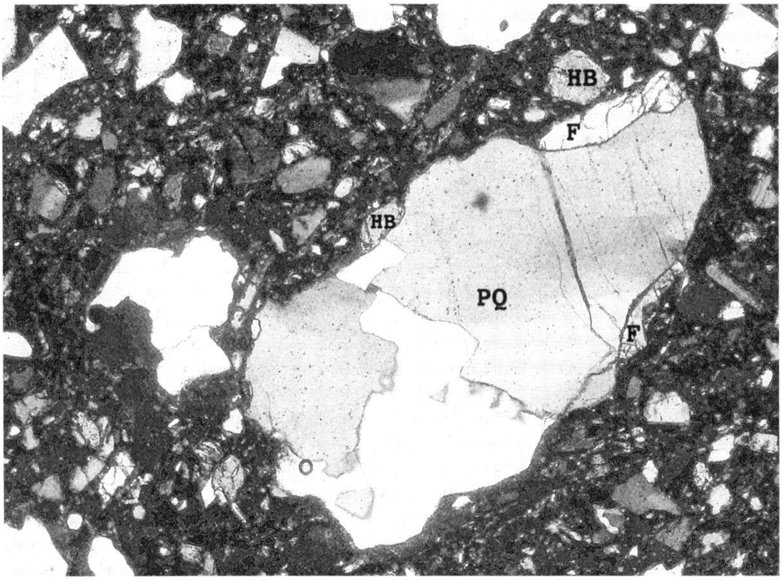

B

5.1. Photomicrograph of local Connestee Simple-Stamped vessel 31-7 from Garden Creek Mound 2 taken at 10X magnification. [A] taken under plane-polarized light, [B] taken under crossed polars. The large granitoid grain is composed of polycrystalline quartz [PQ], slightly altered feldspar [F] (on top and bottom edges at the east end), and hornblende [Hb]. Note the presence of grains of green [under PPL] hornblende (a member of the amphibole family) scattered throughout the vessel body. Scale: large polymineralic grain=1.675 mm.

Table 5.2. Compositional Data for Connestee-Series Vessels from Garden Creek Mound 2

Values in bold fall outside the two-sigma/95% confidence limits for local vessels shown in the shaded cells.

		Bulk Composition				*Mineralogy*		
Type	**Thin Section No.**	**% Matrix**	**% Silt**	**% Sand**	**Sand Size Index**	**% Monoc Quartz**	**% Polyc**	**% FMM**
Local Vessels								
Connestee Simple Stamp	31-1	62	5	33	1.81	15	31	54
Connestee Simple Stamp	31-2	58	7	35	**2.26**	11	63	26
Connestee Simple Stamp	31-3	55	7	38	1.97	8	34	58
Connestee Simple Stamp	31-4	56	7	37	1.78	13	49	38
Connestee Simple Stamp	31-6	52	9	39	1.92	15	67	18
Connestee Simple Stamp	31-7	49	9	42	1.80	14	41	45
Connestee Simple Stamp	31-8	57	4	39	1.96	**20**	44	36
Connestee Simple Stamp	31-9	60	3	37	1.89	12	40	48
Connestee Simple Stamp	31-10	68	4	28	1.75	10	36	54
Connestee Plain	31-13	66	7	27	2.06	10	63	27
Connestee Cordmarked	31-5	62	6	32	2.15	16	68	16
Connestee Cordmarked	31-14	68	6	26	1.82	12	52	36
Mean ±1 Std Deviation	[n=12]	59.4±6.1	6.2±1.9	34.4±5.2	1.93±.16	13.0±3.2	49.0±13.4	38.0±14.2
Two-Sigma Ranges	**[n=12]**	47.4–71.6	2.4–10.0	24.0–44.8	1.61–2.25	6.6–19.4	22.2–75.8	9.6–66.4
Atypical Vessels								
Connestee Simple Stamp	31-11	57	7	36	**1.52**	**51**	28	21
Connestee Simple Stamp	31-12	57	7	36	1.66	**43**	**11**	46
Connestee Simple Stamp	31-15	61	9	30	**1.48**	**22**	**20**	58

Table 5.3. Compositional Data for "Hopewell" and Other Stylistically Nonlocal Vessels from Garden Creek Mound 2

Values in bold fall outside ± 2 standard deviations/95% confidence limits for local vessels shown in the shaded cells.

Type	Thin Section No.	Bulk Composition % Matrix	% Silt	% Sand	Sand Size Index	Mineralogy % Monoc Quartz	% Polyc	% FMM
Hopewell Series Vessels								
Plain Rocker Stamped	31-16	60	6	34	2.03	**6**	58	36
Dentate Rocker Stamped	31-17	61	5	34	2.09	12	70	18
Dentate Rocker Stamped	31-18	63	5	32	2.03	**4**	**78**	18
Zoned Dentate Rocker Stamped	31-19	67	8	25	2.25	17	72	11
Zoned Dentate Rocker Stamped	31-21	61	5	34	2.01	**21**	61	18
Plain Rocker Stamped	31-22	56	5	39	1.90	14	47	39
Dentate Rocker Stamped	31-23	61	6	33	2.14	11	75	14
Cross-Hatched Rim	31-20	66	6	28	**2.68**	**24**	72	**4**
Mean ±1 Std Deviation	**[n=8]**	61.8±3.5	5.8±1.0	32.4±4.2	2.14±.24	13.6±6.9	66.6±10.4	19.8±12.0
Local Two-Sigma Range	**[n=12]**	47.4–71.6	2.4–10.0	24.0–44.8	1.61–2.25	6.6–19.4	22.2–75.8	9.6–66.4
Atypical Vessels								
Turner Simple-Stamped	31-24	56	**11**	33	**1.14**	**47**	35	18
Turner Simple-Stamped	31-25	63	10	27	**1.11**	**42**	57	**1**
Crystal River Plain	31-26	**92**	7	**1**	2.00	**0**	**100**	**0**

Note: St=Standard

5.2. Eight Hopewell-series sherds (with dentate or rocker stamping + one crosshatched rim) from Garden Creek Mound 2. All were thin sectioned, with the appropriate thin-section numbers next to each sherd, and all appear to have been made of local materials.

5.2). The crosshatched rim (31-20; Keel 1976:Plate 18k, right) is a bit different, but considering its rich suite of Appalachian minerals (e.g., epidote, clinozoicite, rutile, and green amphiboles), its differences are best viewed as expressions of local variability rather than evidence of an exotic origin. There are four other instances in Table 5.3 of mineralogical values falling outside the two-sigma range for the 12 local vessels, but all of these departures are minor and are most likely attributable to sampling error within a small sample. From these data, then, it appears that the majority of the Hopewell-style vessels at Garden Creek Mound 2 were made locally (e.g., Figure 5.3). None of those that were thin sectioned can be accepted as an import from Ohio.

Returning to consider the six vessels identified as "Atypical" in Tables 5.2 and 5.3, four can be seen to have much in common. These vessels, 31-11 and 31-12 in Table 5.2 and 31-24 and 31-25 in Table 5.3, are most prominently distinguished from the 20 vessels recognized as "local" (now including the eight Hopewell-series vessels) in possessing a higher incidence of monocrystalline quartz (42–51 percent versus a maximum of 24 percent for "local" vessels of both series; see Tables 5.2 and 5.3). The normal concomitant

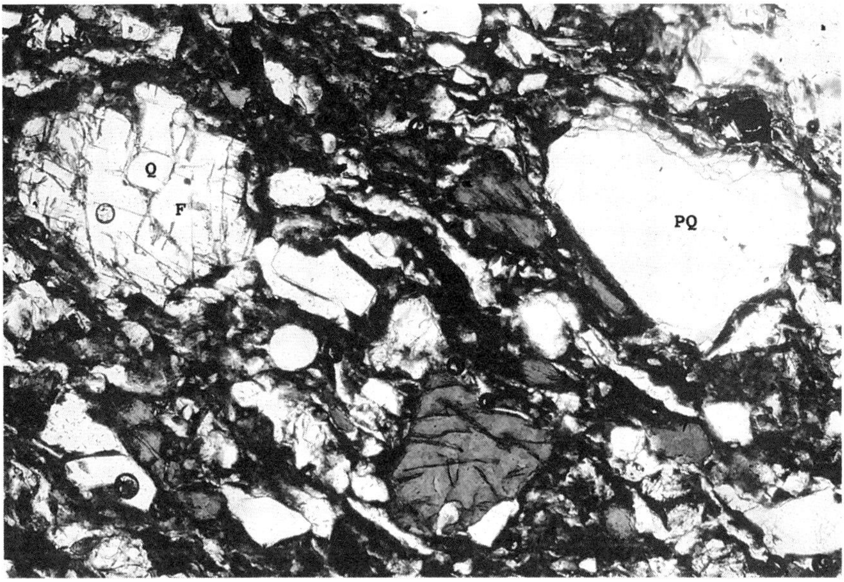

5.3. Photomicrograph of rocker-stamped Vessel 31-16 from Garden Creek Mound 2 taken at 10X magnification under plane-polarized light. Note common scattered green hornblende grains + granitoid grain on left composed of altered feldspar [F] and quartz [Q]. Scale: large polycrystalline quartz grain (upper right, PQ)= .75 mm. Largest hornblende grain (bottom center)= .50 mm. This vessel is considered a local product.

of high monocrystalline quartz content is relatively small sand-size indices, which also characterize these vessels. Variable amounts of green amphiboles occur in all four of these vessels along with clinozoicite, another typical Appalachian mineral, which leads to the acceptance of all four vessels as quartz-rich variants of locally made pottery.

The remaining two Atypical vessels deserve special consideration. Vessel 31-26, which was classified as Crystal River Plain by Keel (1976:118 & Plate 18c), has been discussed in Chapter 3. It most definitely is an import to Garden Creek, almost certainly derived from the Mann site (see discussion in Chapter 3, Table 3.8, and Figure 3.3). The last of these Atypical vessels, 31-15, belongs to a group of distinctive vessels (at least 20 in the present study) that are widespread in Middle Woodland sites from central Ohio to southeastern Tennessee. At the present time a source for these vessels is unknown, but there is little doubt that it is in the Appalachian Mountains. These vessels, assigned to an informal category, "Grit A," will be discussed in Chapter 11.

5.4. A sample of six Connestee Simple-Stamped sherds from Icehouse Bottom. All were thin sectioned, with the appropriate thin-section numbers next to each sherd. Nos. 40-11 and 40-12 are suspected imports from Garden Creek.

Icehouse Bottom

The Icehouse Bottom site is located on a terrace of the Little Tennessee River in Monroe County, southeastern Tennessee (Chapman 1973:Figure 1; Figure 3.1). Jefferson Chapman, then a graduate student at the University of North Carolina, conducted excavations there in 1970 and 1971 then paid a return visit in 1977 (Chapman and Keel 1979:159). From the Middle Woodland component on the site, 29 Hopewell-style sherds, derived from 21 vessels, were recovered along with over 5,000 sand-tempered, Connestee-series sherds and over 15,000 limestone-tempered, Candy Creek-series sherds (Chapman 1973:47–61). Chapman considers the mix of Connestee ceramics and evidence of Hopewell interaction at Icehouse Bottom to be unique in southeast Tennessee (Chapman and Keel 1979:159), where most Middle woodland sites are dominated by Candy Creek pottery. A sample of Connestee Simple-Stamped sherds is presented in Figure 5.4.

A major goal of Chapman's excavations was to establish the stratigraphic/temporal relationship between the Connestee and Candy Creek ceramic series. The "excavations confirmed the contemporaneity of these two series

Table 5.4. Frequencies of Thin-Sectioned Vessels by Ceramic Series from Icehouse Bottom

Ceramic Series/Types	Frequencies
Connestee Series	[13]
Simple Stamped	8
Cordmarked	2
Plain	3
Candy Creek Series	[13]
Bluff Creek Simple-Stamped	9
Candy Creek Cordmarked	2
Wright Check-Stamped	2
Hopewell Series	[3]
Rocker Stamped	3
Totals	29

throughout stratum II" (Chapman 1973:132). It is for this reason that sherds from both series, plus three from the Hopewell series, were selected for thin sectioning. A total of 29 sherds were obtained for thin sectioning in 1985–1986 through the good offices of Jefferson Chapman, who was then director of the Frank H. McClung Museum at the University of Tennessee. Table 5.4 provides a breakdown of those sherds by ceramic type and series.

Again, the first step was to establish a compositional signature for the locally made pottery to serve as a standard for identifying possibly intrusive vessels. Because of the fine-grained makeup of these thin sections, bulk composition and the mineralogical index were adopted to characterize them. Since the limestone-tempered Candy Creek series will be considered in Chapter 8, sherds of this series will not be discussed here. A general review of the local Connestee thin sections identified four that were atypical and considered potentially of nonlocal manufacture (Table 5.5). These four vessels will be considered later in this chapter.

Based upon the initial review, nine stylistically local Connestee vessels appeared to be compositionally homogeneous enough to serve as a reasonably reliable sample for defining the composition of locally made pottery at Icehouse Bottom. Indeed, as can be seen from Table 5.5, these nine vessels are so compositionally uniform that not a single value falls outside of the two-sigma ranges for the group.

Table 5.5. Bulk and Mineralogical Compositional Data for Connestee- and Hopewell-Series Vessels from Icehouse Bottom

Values in bold fall outside ± two standard deviations/95% confidence limits for local vessels shown in the shaded cells.

		Bulk Composition				Mineralogy		
Type	**Thin Section No.**	**% Matrix**	**% Silt**	**% Sand**	**Sand Size Index**	**% Monoc Quartz**	**% Polyc**	**% FMM**
Connestee Series Vessels								
Simple Stamped	40-3	63	13	24	1.40	6	19	75
Simple Stamped	40-4	53	16	31	1.26	12	1	87
Simple Stamped	40-5	67	12	21	1.26	11	0	89
Simple Stamped	40-9	74	7	19	1.53	0	12	88
Simple Stamped	40-10	61	10	29	1.27	2	0	98
Simple Stamped	40-13	63	12	25	1.33	12	7	81
Plain	40-15	64	10	26	1.41	8	10	82
Cordmarked	40-17	59	9	32	1.64	0	0	100
Cordmarked	40-18	74	5	21	1.11	0	0	100
Mean ±1 Std Deviation	**[n=9]**	64.2±6.8	10.5±3.3	25.3±4.6	1.36±.16	5.7±5.3	5.4±7.0	88.9±8.9
Local Two-Sigma Range	**[n=9]**	50.6-77.8	3.9–17.1	16.1–34.2	1.04–1.68	0–16.3	0–19.4	71.1–100
Atypical Vessels								
Simple Stamped	40-11	58	**3**	**39**	1.66	10	**52**	**38**
Simple Stamped	40-12	64	**3**	33	**2.04**	**19**	**61**	**20**

Plain	40-14	63	7	30	1.32	**22**	12	**66**
Plain	40-16	64	**3**	33	**1.76**	3	**28**	**69**
Hopewell-Series Vessels								
Rocker Stamped	40-6*	71	6	23	**3.01**	3	**87**	**10**
Rocker Stamped	40-7	72	7	21	1.51	**60**	**34**	**6**
Rocker Stamped	40-8*	73	5	22	**2.21**	**18**	**62**	**20**

**For bulk composition, temper of grit-tempered vessels is recorded as sand.*

5.5. Photomicrograph of local Connestee Simple-Stamped Vessel 40-3 from Icehouse Bottom taken at 10X magnification under crossed polars. Note abundant clinozoicite grains [Cz] with high relief and anomalous (blue-gray) interference colors. Large chert grain (lower left in between two Cz grains) measures .50 mm.

Overall, the distinguishing compositional properties of these vessels are as follows (refer to Table 5.5 and Figure 5.5): (1) a relatively high incidence of sand-size grains (a mean of 25 percent); (2) a predominance of sand grains that are fine in size (mean sand size index=1.36) and monomineralic in character; and (3) the epidote-family mineral, clinozoicite, is unusually abundant and (along with weakly colored amphiboles) is the main reason for the extremely high FMM percentages that characterize these vessels. Chert, garnet, and rutile are notable, recurrent minor inclusions, while twined feldspars (e.g., plagioclase and microcline) are virtually absent. Chert is an interesting "diagnostic" in that it is ubiquitous in nature but, with the exception of Icehouse Bottom, rarely turns up in Southeastern pottery. For example, it is present in every one of the nine local vessels at Icehouse Bottom (Figure 5.5) yet appears in only one of the 12 local vessels at Garden Creek.

Based upon the establishment of this compositional signature for locally made pottery, the four Connestee-series vessels identified as Atypical do, indeed, appear to be compositionally distinct. Two of these vessels, 40-14 and 40-16, bear the unmistakable fingerprints of Grit A, and will be considered in Chapter 10 with other members of that group. The other two Atypical vessels are considered probable imports from Garden Creek. Table 5.6 com-

Table 5.6. Bulk Composition and Mineralogy of Two Vessels from Icehouse Bottom Believed to Be Derived from Garden Creek Compared with the Two-Sigma Ranges for Local Connestee Vessels at Garden Creek

Type	Thin Section No.	*Bulk Composition* % Matrix	% Silt	% Sand	Sand Size Index	*Mineralogy* % Monoc Quartz	% Polyc	% FMM
Icehouse Bottom								
Connestee Simple Stamped	40-11	58	3	39	1.66	10	52	38
Connestee Simple Stamped	40-12	64	3	33	2.04	19	61	20
Garden Creek								
Two-Sigma Ranges for Local Connestee Series	**[n=12]**	47.4–71.6	2.4–10.0	24.0–44.8	1.61–2.25	6.6–19.4	22.2–75.8	9.6–66.4

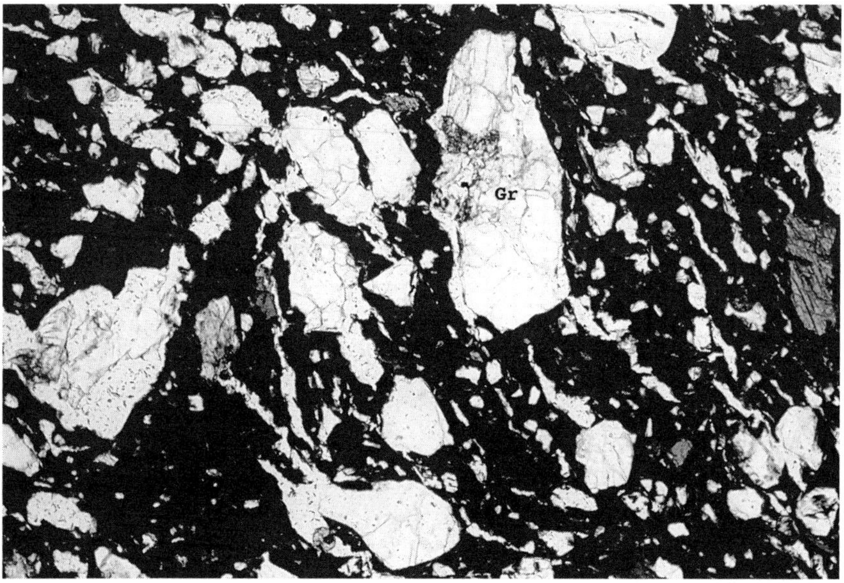

5.6. Photomicrograph of Connestee Simple-Stamped Vessel 40-12 from Icehouse Bottom taken at 10X magnification under plane polars. Large hornblende on right= .35 mm; the 5 other large grains are composed of polycrystalline quartz with minor amounts of altered feldspar [Granitic rock, Gr]. Note also the scattered grains of green-to-brown amphibole. This vessel is a possible intrusive from Garden Creek.

pares the bulk and mineralogy values of these two vessels with the two-sigma range for the 12 local Connestee vessels from Garden Creek. As can be seen from this table, all compositional values for the two Icehouse Bottom vessels fall fully within the two-sigma range for local Garden Creek pottery, thus supporting the view of their importation from the Garden Greek region if not the site itself (Figure 5.6).

Likewise, two of the rocker-stamped vessels appear to be nonlocal in character, both stylistically and compositionally (see Figure 2.11 and Table 5.5). In an earlier report (Stoltman 1999b), I identified two of the three rocker-stamped vessels (see Nos. 40-6 and 40-8 in Table 5.5) as imports from Ohio, while the third vessel (40-7) was suspected of being an atypical, but local, product. The quartz-dominated paste of the latter (see Table 5.5) is unique at the site, while the presence of minor amounts of clinozoicite, green amphiboles, and chert are consistent with Icehouse Bottom clays (Figure 5.7). The most parsimonious interpretation of these data is that vessel 40-7, which is sand, not grit-tempered, is an atypical local product since no external source for it is known.

5.7. Photomicrograph of sand-tempered sherd [mostly quartz] no. 40-7, probably of local origin, from Icehouse Bottom taken at 10X magnification under crossed polars. Scale: clinozoicite (just above center with high relief)=.175 mm.

5.8. Photomicrograph of Simple-Stamped Vessel 33-193 recovered at Seip and believed to be derived from Icehouse Bottom (Cf. figure 5.5). Image taken at 10X magnification under crossed polars. Note the scattered clinozoicite [Cz] grains (largest on far left measures .275 mm. Large chert [Ch] upper right=.50 mm.

Table 5.7. Bulk Composition and Mineralogy of Two Connestee Simple-Stamped Vessels from Seip Believed to Be Derived from Icehouse Bottom Compared with the Two-Sigma Ranges for Connestee-Series Vessels from Icehouse Bottom

Type	Thin Section No.	*Bulk Composition* % Matrix	% Silt	% Sand	Sand Size Index	*Mineralogy* % Monoc Quartz	% Polyc	% FMM
Seip								
Connestee Simple Stamped	33-184	56	11	33	1.33	10	10	80
Connestee Simple Stamped	33-193	65	12	23	1.50	8	16	76
Icehouse Bottom								
Two-Sigma Ranges for Local Connestee Series	**(n=9)**	50.6–77.8	3.9–17.1	16.1–34.2	1.04–1.68	0–16.3	0–19.4	71.1–100

As for vessels 40-6 and 40-8, their Ohio origins cannot be doubted (see Table 2.22). The presence of granitic (grit) temper accompanied in one case (vessel 40-8) by shale are distinctive Ohio properties (see Figure 2.12). The new compositional data presented in Table 5.5 provide further confirmation.

In addition, there is firm evidence from the Seip site that ceramics moved in both directions between Ohio and Tennessee. Table 5.7 records the compositional data for two simple-stamped vessels recovered at Seip along with the two-sigma ranges for the local Connestee-series vessels from Icehouse Bottom. As can be seen from Table 5.7, all compositional values for these two vessels fall within the two-sigma range for local pottery from Icehouse Bottom. Not only quantitatively but also qualitatively, these vessels unmistakably bear the impress of the Icehouse Bottom region, even possessing chert along with the characteristic abundance of clinozoicite and other Appalachian minerals such as rutile and garnet (Figure 5.8).

6

The Appalachian Plateau and Ridge and Valley Provinces of Northwest Georgia

Two important Middle Woodland sites located in the northwest corner of Georgia are included in this study: Tunacunnhee and Leake (see Figure 3.1). Each will be discussed in turn.

Tunacunnhee

The Tunacunnhee site is a Middle Woodland mound group (comprised of eight mounds) and associated habitation area located in a small river valley on the Cumberland Plateau in the extreme northwest corner of Georgia (Jefferies 1976:1–2; 1979). Excavations directed by Joseph Caldwell, in behalf of the University of Georgia, were conducted in 1973. The focus was upon the mounds (all eight were tested, four proving to be modern) with only "a small test excavation" in the habitation area (Jefferies 1976:5). The preponderance of the artifacts recovered came from major excavations of four of the mounds, A, C, D, and E, all of which had generous amounts of limestone slabs in their construction. The archaeological assemblage is uniform enough to be considered derived from a single component.

The site is noteworthy for "the largest concentration of Hopewellian items in the interior southeast" (Jefferies 1976:48). Included among them, all found in mortuary contexts, are a rich array of copper artifacts, including panpipes, bicymbal earspools, and plates, as well as a variety of other items typically associated with Hopewell interaction such as mica plates and cutouts, platform pipes, bear canines, and shark teeth (Jefferies 1976: 21–29; see also Jefferies 2006). Overall, the recovery of pottery was sparse (1,693 sherds in all; Jefferies 2006:167), with no vessel found as a grave good, while "ceramic artifacts were almost totally absent from the mound fill" (Jefferies 1976:31). From the habitation area, 523 sherds were recovered (Jefferies 1976: 32).

The ceramics analyzed for this study include 20 sherds derived from both midden and feature contexts kindly provided by Professor David Hally of the

Table 6.1. Frequencies of Thin-Sectioned Vessels by Temper Group and Surface Treatment from the Tunacunnhee Site

Temper Classes	Simple Stamped	Plain	Cordmarked	Totals
"Grit 1"=Fissured Granite	7	1		8
"Grit 2"=Metaquartzite		4		4
Limestone	2	1	4	7
Unique Chert tempered	1			1
Totals	10	6	4	20

University of Georgia. Table 6.1 records the frequencies of sherds by temper class and surface treatment. The initial screening of the Tunacunnhee thin sections revealed that 13 were tempered with a variety of rock fragments while seven were limestone tempered. The seven limestone-tempered sherds were set aside for consideration in Chapter 8.

From the initial screening of the 13 thin sections, it was decided that distinguishing fine grit from natural sand could not be accomplished objectively, so the bulk composition index was used along with mineralogical identifications. These observations are recorded in Table 6.2. As can be seen from this table, two kinds of fine grit were recognized, one derived from a granitic rock (n=8) and the other from a metaquartzite (n=4). One vessel (9-8) is unique in possessing a high incidence of chert and, indeed, can be considered chert tempered. It appears to be an import but from an unknown source.

Two distinctive tempers, informally referred to as Grit 1 and Grit 2, are recognizable among the 12 vessels presumably locally made at Tunacunnhee. Both were probably derived from relatively finely crushed metamorphic rocks. Grit 1 is derived from a granitic rock in which quartz occurs in highly variable amounts (mostly monocrystalline and often coarse-grained), along with feldspars (either untwined or plagioclase), amphiboles, both muscovite and biotite micas, and scattered clinozoicite. The most distinguishing feature is the prominence of "fissured feldspars," which are sometimes directly associated with quartz in coarse polycrystalline grains (Figure 6.1). Most common, however, are monocrystalline feldspar grains, which are the main reason for the elevated FMM content (see Table 6.2). The term "fissured" pertains to the distinctive chemical alteration (brown or dark discoloration) that occurs along fissures penetrating into the feldspar grains (see Figure 6.1). While the suite of minerals is the same for all Grit 1 vessels, two minerals in par-

Table 6.2. Bulk Composition and Mineralogy of Local Grit-Tempered Vessels from Tunacunnhee

		Bulk Composition				*Mineralogy*		
Type	**Thin Section No.**	**% Matrix**	**% Silt**	**% Sand**	**Sand Size Index**	**% Monoc Quartz**	**% Polyc**	**% FMM**
"Grit 1"- Fissured Granite								
Simple Stamped	9-1	64	4	32	1.95	22	35	43
Simple Stamped	9-4	61	3	36	1.96	26	15	59
Simple Stamped	9-5	67	1	32	2.13	6	16	78
Simple Stamped	9-6	63	1	36	1.79	4	4	92
Simple Stamped	9-7	67	3	30	1.85	4	11	85
Plain Foot	9-9	67	3	30	1.93	51	9	40
Simple Stamped	9-12	58	6	36	1.93	23	21	56
Simple Stamped	9-19	64	1	35	2.12	51	27	22
Mean ±1 Std Deviation	**(n=8)**	63.9±3.2	2.7±1.8	33.4±2.7	1.96±.12	23.4±19.2	17.2±10.1	59.4±24.3
"Grit 2"- Metaquartzite								
Plain	9-2	73	3	24	2.87	2	71	27
Plain	9-3	63	5	28	2.52	4	76	20
Plain	9-13	67	5	28	2.81	0	85	15
Plain	9-14	70	2	28	3.00	2	87	11
Mean ±1 Std Deviation	(n=4)	68.2±4.3	3.8±1.5	28.0±3.3	2.80±.20	2.0±1.6	79.8±7.5	18.2±6.9
Chert-Tempered, Simple Stamped	9-8	91	3	6	3.00	0	93	7

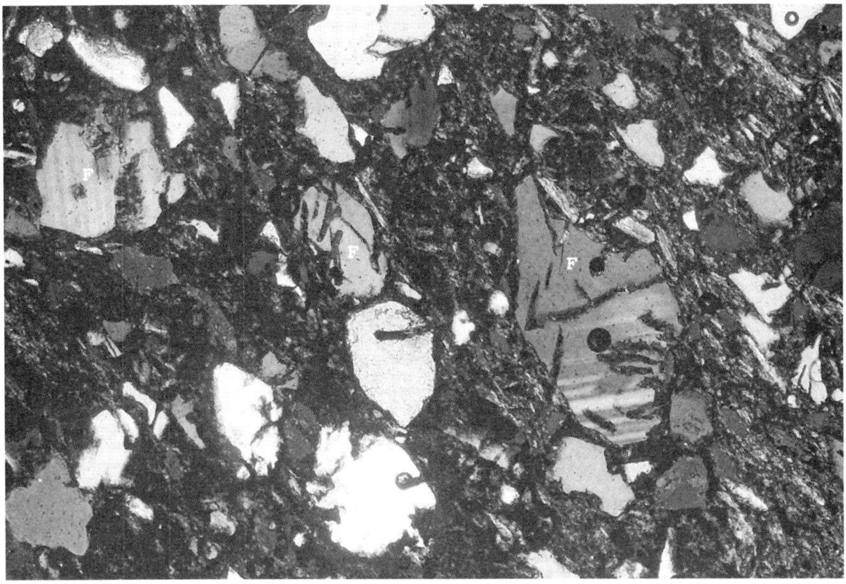

6.1. Photomicrograph of Simple-Stamped Vessel 9-4 from Tunacunnhee showing multiple grains of fissured feldspars [F]. Image taken at 10X magnification under crossed polars. Scale: large fissured and twinned plagioclase right center= .775 mm.

ticular, quartz and amphiboles, are noteworthy for the great variation in their frequencies: quartz percentages range from 4 percent to 51 percent of sand, while amphibole percentages range from trace amounts to 26 percent of sand.

By contrast, polycrystalline quartz grains dominate Grit 2 (note the high values for percent Polyc in Table 6.2). That this is a metamorphic rock (metaquartzite) is evidenced by the frequent deformation of the quartz crystals, their undulose extinction, and sutured grain boundaries (Figure 6.2). What is especially distinctive about these thin sections, beside the quartzite, is the occurrence of heavily altered (but rarely fissured) feldspars. These feldspars sometimes occur in association with the polycrystalline quartz but they more commonly occur as loose monocrystalline grains that are so heavily altered that they appear dark brown and nearly amorphous (see Figure 6.2, upper right). Biotite and amphiboles are sparsely represented, but clinozoicite seems to be absent.

Leake

The Leake site (Keith 2010a) is located on the Etowah River in the Great Valley of the Ridge and Valley Province just inside the city limits of Cartersville,

6.2. Photomicrograph of undecorated vessel 9-13 from Tunacunnhee taken at 10X magnification under crossed polars. Note several grains of "Grit 2," metaquartzite [MQ], plus altered feldspar upper right [F]. Scale: altered feldspar= .45 mm; largest metaquartzite left center=1.05 mm.

Bartow County, Georgia (see Figure 3.1). Excavations, which were designed to mitigate the effects of the widening of State Highways 61 and 113, were conducted from November 2004 through September 2005, with additional work done in December 2006. Some 50,000 square feet were excavated, exposing extensive Middle Woodland occupation that covered an estimated 500 acres and included 3 mounds, a circular ditch, extensive midden deposits often reaching over 40 cm in thickness, and thousands of subsurface features.

There are two major Middle Woodland components (Keith 2010a:59). The earliest, referred to as Cartersville, is characterized by the predominance of Cartersville Check Stamped pottery along with lesser amounts of plain and Dunlap Fabric Impressed vessels. The younger component, referred to as Swift Creek, has a ceramic assemblage dominated by Swift Creek Complicated Stamped (63.9 percent by weight), Cartersville Simple Stamped (24.5 percent by weight), and Cartersville Check Stamped (9.7 percent) (Keith 2010a:307). Several unusual minority types are also present including diamond check stamped, red slipped, and incised and trailed vessels, some with applique features. Most distinctive of the stylistically anomalous vessels are several that closely resemble Lower Valley types, especially Churupa Punc-

6.3. Selected sherds from the Leake site showing the stylistic diversity present at the site. A=Cartersville Simple Stamped; B=Cartersville Check Stamped; C=Swift Creek Complicated Stamped; vessel 9-270 is a probable import from the Coastal Plain; D=Churupa Punctated-like sherds.

tated (cf. Figure 6.3D and Phillips 1970:67–68), yet all are made of local materials. It is important to note that no other Middle Woodland component in the entire Southeast has such ceramic diversity (Figure 6.3). My contribution to this project involved conducting petrographic analyses of 134 sherds and one clay pipe. My report appears as Appendix 1 in Volume II of the final report (Keith 2010a; Stoltman 2010).

It is in the Swift Creek component that evidence of participation in the Hopewell Interaction sphere occurs, including prismatic blades of Flint Ridge chert from Ohio, cut mica plates, and fragments of copper plates. However, such items are neither as abundant nor as dramatic as at Tunacunnhee, no doubt because no mortuary features were encountered (there was only limited testing in one of the mounds).

Table 6.3 records the frequency of thin-sectioned sherds by typological class, excluding limestone-tempered and grog-tempered vessels that will be considered later. Note that there are eight typological classes in marked contrast to Tunacunnhee, located barely 100 km to the northwest, that has only three ceramic types, plain, simple-stamped, and cordmarked. Table 6.3 records the bulk compositional and mineralogical mean values for vessels in each of the typological classes plus the pipe.

The initial review of these thin sections, as is common for Southeastern pottery generally, revealed them to be too fine grained to permit reliable discrimination between natural sands and temper. Accordingly, bulk composition and the mineralogical values were adopted as the most appropriate means to characterize their compositions. It can be seen from Table 6.3 that bulk composition is surprisingly uniform among the typological classes, with matrix percentages averaging between 64 percent and 71 percent, silt percentages averaging between 5.6 percent and 9.5 percent, and sand percentages averaging (with the exception of the pipe) between 21 percent and 30 percent. Moreover, the standard deviations are relatively small (all but one below 10 percent) indicating relative uniformity within each of the classes. Especially noteworthy are the small sand size indices, which range within the narrow limits of 1.53 to 1.83 (again excepting the pipe). One could reasonably conclude from these data that the uniformity in bulk composition among the typological classes is consistent with the view that all were made locally following similar practices of processing raw materials.

The initial review of the mineralogical makeup of the Leake thin sections, after temporarily setting aside the seven limestone-tempered and two grog-tempered vessels (discussed in Chapter 3; Table 3.8), revealed a wide range of diversity within a limited set of rocks and minerals. A general suite of minerals was almost always present, with monocrystalline grains of the following minerals virtually ubiquitous: quartz, plagioclase, untwined feldspar (which could be either orthoclase or plagioclase), various members of the amphibole family, muscovite, and clinozoicite. Polymineralic rock fragments also occur, but usually in minority amounts. The rock fragments observed, with rare exceptions, were derived from metamorphic rocks of either granitic or quartzitic composition. The main (sometimes only) constituent of such rocks is polycrystalline quartz that shows unambiguous evidence of metamorphic alteration, such as undulose extinction, sutured grain boundaries, annealed structure (triple-junction grain intersections), and frequent signs of mechanical deformation (mylonization). Sometimes the polycrystalline quartz is associated with other minerals, principally feldspar (both twinned and untwinned) that shows evidence of chemical alteration, visible as brown discoloration along "fissures" or even nearly total brown discoloration. Less commonly associated with the polycrystalline quartz are amphiboles, muscovite, and clinozoicite. Whether one rock (an impure metaquartzite) or two separate rocks (metaquartzite and metagranite) are represented is uncertain.

Table 6.3. Means ± One Standard Deviation for Bulk and Mineralogical Composition of Leake Pottery by Typological Class

Seven limestone-tempered and two grog-tempered vessels not included.

		Bulk Composition				*Mineralogy*		
Type	**N**	% **Matrix**	% **Silt**	% **Sand**	**Sand Size Index**	**% Monoc Quartz**	% **Polyc**	% **FMM**
Swift Creek Comp Stamped	47	66.4±6.5	8.3±4.5	25.3±7.5	1.63±.26	43.0±19.4	31.5±13.6	25.5±17.2
Diamond Check Stamped	18	64.2±10.0	6.0±2.2	29.8±9.4	1.83±.27	31.0±22.6	37.2±14.6	31.8±18.0
Check Stamped	19	66.3±6.5	5.6±2.5	28.1±6.4	1.75±.26	31.2±16.5	38.0±17.2	30.8±19.9
Simple Stamped	6	68.0±6.8	9.5±5.5	22.5±3.8	1.53±.16	35.0±23.2	22.8±12.5	42.2±17.5
Incised and Trailed	19	68.7±5.7	7.0±5.3	24.3±7.4	1.59±.30	41.2±19.9	25.8±13.8	33.0±17.1
Red Slipped	3	70.0±1.7	8.3±5.1	21.7±6.8	1.83±.18	46.3±19.1	21.0±10.4	32.7±13.4
Fabricmarked/Cordmarked	8	67.0±4.2	8.0±2.6	25.0±4.1	1.73±.23	33.1±19.9	31.8±15.0	35.1±11.6
Plain	5	71.0±6.2	7.4±7.3	21.6±6.6	1.70±.35	49.2±16.8	23.4±13.6	27.4±16.2
Pipe	1	77	7	16	2.01	14	73	13
Total Thin Sections	126							

The diversity among the three mineralogical classes presented in Table 6.3 is much greater than characterized the bulk compositional data. The most striking features of these data are the large standard deviations—all are in double digits—indicating an enormous amount of within-class variability. The lowest mean percent for a mineralogical class is 21 percent, the highest 49.2 percent, so that with such large standard deviations, it is virtually impossible to see evidence of any notable covariation between a typological class and a mineralogical class. In an effort to make more sense of the compositional data, each of the 126 thin sections characterized by fine grit were assigned to one of six petrographic classes. These petrographic classes were based upon subjective criteria that were thought best to differentiate the thin sections into groups that were compositionally distinct and recognizable. The petrographic classes, each named after the prevailing or most characteristic inclusions, are described below. (Thin sections of vessels from four of these classes are portrayed in Figure 6.4.)

(1) Monocrystalline Quartz=grains of monocrystalline quartz constitute a minimum of 50 percent of those counted (see Figure 6.4A).
(2) Metagranite=the co-occurrence of polycrystalline quartz and chemically altered feldspars in single grains; because of the fineness generally of all mineral inclusions, the altered feldspar often occurs in monocrystalline grains that attain frequencies of 40–50 percent of all inclusions (see Figure 6.4B).
(3) Metaquartzite=greater than 50 percent of inclusions are polycrystalline quartz grains with metamorphic properties.
(4) Amphibole-rich=whenever the FMM group constitutes greater than 40 percent of total inclusions and amphiboles stand out as the most predominant member of the group (see Figure 6.4C).
(5) Composite=frankly a residue after thin sections had been assigned to the above four mineralogical classes (see Figure 6.4D).
(6) Opaque-rich=a unique, simple-stamped vessel with an opaque-rich temper that can be assigned to no other class.

Table 6.4 records the bulk composition and mineralogy for each of these petrographic classes. As with the typological classes of Table 6.3, the bulk compositions among the petrographic classes in Table 6.4 show considerable uniformity, including relatively small standard deviations that do not exceed ±8.3. That the mineralogical classes are more variable between classes is no surprise since they were defined to yield compositional patterning. For example, it comes as no surprise that the Monocrystalline Quartz dominated class is characterized by an average of 63% quartz, double that all but one of

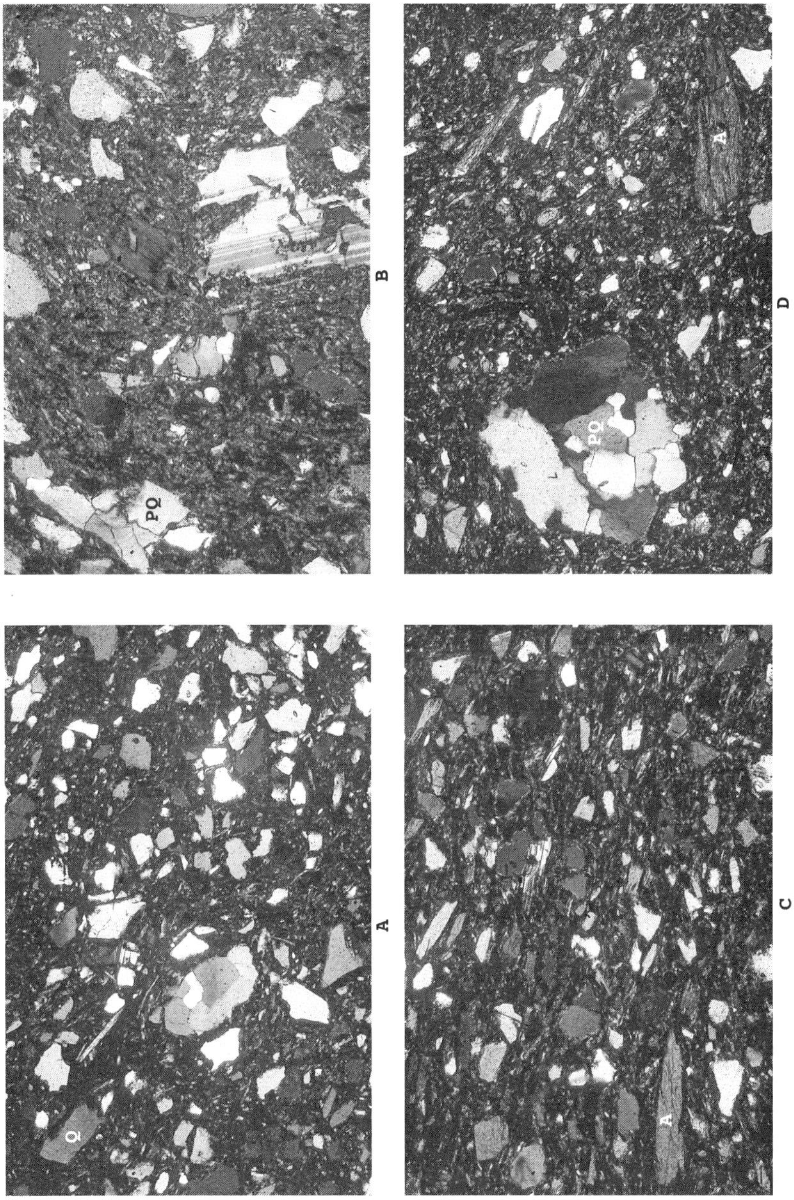

6.4. Photomicrographs of four Simple-Stamped vessels from the Leake site, each representing a different petrographic class: A=monocrystalline quartz dominant (9-154) Quartz grain [Q]=.30mm; B=metagranite (9-215); Polycrystalline quartz grain [PQ]= .95mm; C=amphibole-rich (9-153); Amphibole grain [A]=.68mm; D=composite (9-156); Polycrystalline quartz grain [PQ]=.95mm; Amphibole grain [A]=.90mm. All views taken at 10X magnification under crossed polars.

the other four classes (see Table 6.4). The question then becomes, were these patterns discovered or created?

The data in Table 6.5 address this question. In this table the frequency of vessels per typological class are tabulated versus each of the five main petrographic classes for 124 vessels (The pipe and the seven limestone-tempered, two grog-tempered, and unique opaque-tempered vessels are excluded). The rationale is that if culturally significant compositional variation occurs within the thin section data, there ought to be significant covariation (i.e., nonrandom clustering) within these data. As can be seen from Table 6.5, considering the frequency differences between the various classes, the distribution of vessels among all the classes shows no clustering whatsoever, indeed appears to be random. Stated another way, every ceramic type is characterized by multiple petrographic classes, even the three red-slipped vessels, which fall into three petrographic classes.

The Leake site provides an interesting and important object lesson for the interpretation of petrographic data. A common expectation, certainly one that was held during most of the course of this study, is that locally made pottery at a site would be relatively uniform in composition as a result of the makers applying culturally conditioned and generally shared practices for making pottery using a limited range of local materials. The mineralogical diversity observed within the Leake ceramic assemblage, however, does not conform to this expectation. How to explain this state of affairs?

First, we must place the Leake site in its geological context. It is located in an alluvial valley that drains a geologically complex mountainous region, much of which is underlain by metamorphic rocks. The Leake pastes were certainly derived from alluvium, but whether used "raw" or processed in some way is unknown (no local sediments were thin sectioned). Whatever the derivation of the sediments, Leake pottery is full of mineral inclusions that are ultimately derived from metamorphic rocks, and such rocks are notable for the wide array of minerals that characterize them. In short, Leake potters had ready access, whether from alluvium or bedrock or both, to a mineralogically diverse landscape. It thus should be no surprise that the mineralogy of Leake pottery is highly varied.

How the potters dealt with this mineralogical diversity is an important factor that is difficult to evaluate, but the practices of the potters were certainly important contributors to the composition of the Leake pottery. If my inference is correct that the Leake potters added finely crushed rock fragments (i.e., grit) to their pastes (i.e., did not use only natural sands), the fine texture and great variability of mineral inclusions can be seen as a natural byproduct of crushing and grinding complex metamorphic rocks.

Table 6.4. Means ± One Standard Deviation for Bulk and Mineralogical Composition of Leake Fine-Grit and/or Sand-Tempered Pottery by Petrographic Class

		Bulk Composition				*Mineralogy*		
Petrographic Class	**N**	% **Matrix**	% **Silt**	% **Sand**	**Sand Size Index**	**% Monoc Qtz**	% **Polyc**	% **FMM**
Monocrystalline Qtz Dominance	37	67.8±6.5	6.4±3.9	25.8±7.5	1.62±0.26	62.7±9.2	21.4±9.4	15.9±9.0
Metagranite Prominence	23	63.1±8.3	6.6±2.8	30.8±8.3	1.94±0.24	15.3±6.3	50.8±13.3	33.9±15.1
Amphibole Rich	14	67.7±5.3	6.0±3.6	26.3±6.9	1.56±0.17	17.2±8.6	22.1±9.8	60.7±13.0
Metaquartzite Prominence	4	64.0±8.3	4.8±1.7	31.2±7.1	2.01±0.13	28.5±13.4	56.5±6.5	15.0±10.2
Composite	47	67.8±6.2	9.1±5.0	23.0±6.0	1.61±0.24	37.6±9.1	31.9±9.4	30.5±10.4
Opaque Rich	1	77	7	16	1.70	33	8	59
Total Thin Sections	126							

Note: Qtz=Quartz

Table 6.5. Frequencies of Thin-Sectioned Vessels by Type vs. Petrographic Class at Leake

The pipe and the unique opaque-tempered, simple-stamped vessel are excluded.

Type	**Monocrystalline Quartz Dominance**	**Metagranite**	**Amphibole rich**	**Metaquartzite**	**Composite**	**Totals**
Swift Creek Complicated Stamped	18	5	6	—	18	47
Diamond Check Stamped	4	10	—	1	3	18
Check Stamped	3	4	3	3	6	19
Simple Stamped	1	—	2	—	2	5
Incised and Trailed	7	2	2	—	8	19
Red Slipped	1	—	1	—	1	3
Fabricmarked/Cordmarked	1	1	—	—	6	8
Plain	2	—	—	—	3	5
Totals	37	22	14	4	47	124

In sum, three major conclusions have emerged from this analysis of Leake pottery thin sections: (1) a consistent suite of metamorphic minerals (e.g., monocrystalline and polycrystalline strained quartz, chemically altered feldspars, amphiboles, muscovite, and clinozoicite) characterize them individually as well as collectively; (2) while each of these minerals is normally present, their percentages from thin section to thin section vary almost wildly; and (3) despite the enormous (indeed unparalleled) typological diversity of the Leake ceramics, the compositional evidence strongly suggests that the great majority of vessels were made following a consistent set of *local* practices. In Chapter 10 we shall return to address more specifically the significance of the apparent compositional uniformity among such a stylistically diverse ceramic assemblage.

7
The Coastal Plain of Georgia and Florida

Three sites from this region are included in the present study: Mandeville and Kolomoki in southwest Georgia and Crystal River on the Florida Gulf Coast. All are well-known for their apparent participation in the Hopewell Interaction Sphere. Each will be discussed in turn.

Mandeville (9 Cla 1)

The Mandeville site is located near the floodplain of the Chattahoochee River about five miles upstream from the Walter F. George Lock and Dam near Fort Gaines in southwest Georgia. Today, the site lies beneath 12 m of water in the Walter F. George reservoir (Kellar et al. 1962; Smith 1979).

The University of Georgia under the direction of A. R. Kelly conducted excavations here from 1959 to 1962, with a preliminary report published promptly during the final year of fieldwork (Kellar et al. 1962). The site consisted of two mounds and a village area. The largest mound, Mound A, was a flat-topped platform that had a Mississippian component on top, which buried several layers of Middle Woodland occupation in the lower levels. Mound B was a smaller conical mound that yielded a rich array of Hopewellian mortuary goods, including 5 copper-covered panpipes, 14 bicymbal earspools, a human ceramic figurine, 2 clay platform pipes, cut mica, and several Flint Ridge chert prismatic blades (Kellar et al. 1962:342).

In the preliminary report, the excavators recognized three components that were referred to as Mandeville I, assigned to Deptford culture, Mandeville II, referred to as Swift Creek, and Mandeville III, assigned to Roods, a Mississippian phase (Kellar et al. 1962). Betty Smith, later reanalyzing the ceramics in a doctoral dissertation, citing strong evidence of continuity between the earliest two components in Mound A, concluded that only a single Middle Woodland component was present, which she preferred to call Early

Swift Creek (Smith 1975; 1979:182). The thin sections described herein from Mandeville were derived from this component.

A sample of sherds from ten Mandeville vessels was obtained from David Hally at the University of Georgia for thin section analysis. Table 7.1 itemizes these sherds by type and records their bulk and mineralogical compositions. The usual initial review of the thin sections before point counting was conducted encountered no obvious compositional outliers and, considering their generally fine texture, the use of bulk and mineralogical indices was deemed most appropriate for characterizing the compositions of what are generally considered sand-tempered ceramics.

Using the two-sigma rule (95 percent confidence limits) to identify compositional values that are statistical outliers indicative of possible nonlocal status (see the bottom of Table 7.1), it can be seen that all compositional values except one (percent silt for vessel 9-28) fall within the two-sigma limits for the site. The one exception exceeds the upper two-sigma value by .6 percent, which can surely be attributed to sampling error. The compositional data for the Mandeville sherds are so homogeneous that it is reasonable to accept all as locally made.

The salient property of the Mandeville compositional data is the high incidence of monocrystalline quartz—a mean of 80.4 percent and a minimum of 64 percent. The relatively high percentage of sand (28 percent) and the relatively modest sand-size indices (1.55) are expectable correlates of quartz-dominated ceramic bodies (Table 7.1). Figure 7.1 shows a thin section of a typical Coastal Plain vessel from Mandeville.

Kolomoki (9 Er 1)

The Kolomoki site attained prominence in the literature of Southeastern archaeology in the aftermath of William Sears's 1948–1952 excavations (Sears 1956). The site is located in the Chattahoochee Valley of southeast Georgia less than 200 km south of Mandeville. It is comprised of at least nine mounds, the largest of which is a 17-m high platform. Considerable controversy surrounded Sears's early, vehement advocacy of the belief that Kolomoki was a post-Woodland site coeval with the Mississippian tradition (e.g., Pluckhahn 2003:4–6). In light of recent research "The Kolomoki Problem" (Pluckhahn 2003:4) has been resolved in favor of the view that there were two main pre-Mississippian (i.e., Woodland) periods of occupation at the site. The first, or Kolomoki phase, is a regional variant of Swift Creek culture that dates ca. A.D. 350–500 near the end of the main period of Hopewell interaction (Pluck-

Table 7.1. Bulk and Mineralogical Composition of 10 Vessels from Mandeville (9Cy1)

Value in bold font falls outside the relevant two-sigma limits for the site shown in the shaded cell.

		Bulk Composition				*Mineralogy*		
Type	**Thin Section No.**	**% Matrix**	**% Silt**	**% Sand**	**Sand Size Index**	**% Monoc Quartz**	**% Polyc**	**% FMM**
Simple Stamped	9-25	71	8	21	1.75	77	23	0
Simple Stamped	9-26	56	8	36	1.65	88	9	3
Check Stamped	9-21	75	9	16	1.08	94	6	0
Check Stamped	9-22	60	3	37	1.99	96	4	0
Check Stamped	9-23	62	3	35	1.83	64	11	25
Check Stamped	9-24	62	5	33	1.05	81	3	16
Complicated Stamped	9-27	78	3	19	1.12	80	8	12
Complicated Stamped	9-28	55	**14**	31	1.30	64	22	14
Complicated Stamped	9-29	74	7	19	1.93	76	9	15
Plain	9-30	63	4	33	1.81	84	11	5
Mean ±1 Std Deviation	**[n=10]**	65.6±8.2	6.4±3.5	28.0±8.2	1.55±.37	80.4±10.9	10.6±6.8	9.0±8.6
Two-Sigma Ranges	**[n=10]**	49.2-82.0	0-13.4	11.6-44.4	1.00-2.29	58.6-100	0-24.2	0-26.2

7.1. Photomicrograph of sand-tempered, Simple-Stamped Vessel 9-26 from Mandeville taken at 10X magnification under crossed polars. Typical Coastal Plain vessel with 88 percent monocrystalline quartz. Scale: largest quartz grain [Q] near center= .625 mm.

hahn 2003:17; Steinen 2006:178–181). Evidence of Hopewell interaction at Kolomoki (especially in Mound E) is less specific than at Mandeville and is consistent with the view that Mandeville predates the earliest occupation at Kolomoki (Steinen 2006:180). A ceramic sample from Kolomoki may thus be only marginally relevant to the main theme of this study, but its inclusion, nonetheless, is valuable as a source of further evidence of the compositional signature of southwest Georgia pottery in broadly Middle Woodland contexts.

A sample of sherds from six Swift Creek Complicated Stamped vessels from Kolomoki was available for this study. Following the usual pre-point-count review of the thin sections, there was no doubt of their fine-grained texture and the appropriateness of bulk composition as the most suitable index with which to characterize their compositions.

Table 7.2 records the bulk and mineralogical compositions of these six vessels. The two-sigma range for the six vessels is recorded at the bottom of the table. From these data, it can be seen that not a single value for any of the six vessels falls outside of the relevant 95 percent confidence limits. The vessel compositions are extraordinarily homogeneous, consistent with the view

Table 7.2. Bulk and Mineralogical Composition of Six Vessels from Kolomoki (9Er1)

All values fall within the site's two-sigma ranges/95% confidence limits.

		Bulk Composition				*Mineralogy*		
Type	**Thin Section No.**	**% Matrix**	**% Silt**	**% Sand**	**Sand Size Index**	**% Monoc Quartz**	**% Polyc**	**% FMM**
Swift Creek Comp Stamped	9-70	61	4	35	1.59	86	6	8
Swift Creek Comp Stamped	9-72	66	2	32	1.41	81	12	7
Swift Creek Comp Stamped	9-123	70	7	23	1.11	82	11	7
Swift Creek Comp Stamped	9-124	56	4	40	1.49	86	3	11
Swift Creek Comp Stamped	9-125	75	2	23	1.64	80	10	10
Swift Creek Comp Stamped	9-126	59	3	38	2.15	84	13	3
Mean ±1 Std Deviation	**[n=6]**	64.5±7.2	3.7±1.9	31.8±7.4	1.56±.34	83.2±2.6	9.2±3.9	7.6±2.8
Two-Sigma Range	**[n=6]**	50.1-78.9	0-7.5	17.0-46.6	1.00-2.24	78.0-88.4	1.4-17.0	2.0-13.2

Note: Comp=Complicated

that all were locally made. By comparing the data in Table 7.2 with those in Table 7.1, it is evident that the two samples provide a consistent picture of the composition of Middle Woodland pottery of southwest Georgia, i.e., the clear predominance of monomineralic quartz of fine to medium size grades.

The same compositional pattern is evident among the Swift Creek sites reported by Stoltman and Snow (1998) from the Coastal Plain of south central Georgia. In that study only bulk compositions were recorded (no need to repeat them here), but the addition of the mineralogical composition can be usefully added in the present context. Among 4 sites with a minimum of 6 thin-sections (Milamo, Lind Landing, Teresa Raceway, and Broxton Creek), 30 of 34 vessels have monocrystalline quartz values that range between 70 percent and 100 percent. Three vessels have metagranite tempers and were certainly imported from the highlands, leaving one atypical vessel from Teresa Raceway with only 46 percent monocrystalline quartz inclusions.

Crystal River (8 Ci 1)

Crystal River is the southernmost site with unambiguous evidence of Hopewell interaction. Three visits to the site by C. B. Moore in 1903, 1906, and 1917 (Moore 1903, 1907, and 1918) recovered most of the Hopewellian artifacts for which the site is famous. These artifacts, including bicymbal earspools with silver coating, disks with cutout designs, a repoussé plate, and a panpipe, all of copper (Moore 1903:409–411), were recovered as mortuary accompaniments in the central mound complex of the site. William Sears (1962) assigned these and similar artifact types from two other Florida Gulf Coast sites to the "Yent Complex," a mortuary or ceremonial complex with Hopewellian ties that was part of a larger cultural context that is otherwise native Floridian (Milanich 1994:135).

Subsequent excavations at the site by Ripley Bullen in the 1950s and 1960s (most of which are unpublished, but see Bullen 1953) have considerably broadened our understanding of the site. In contrast to Gordon Willey (1949:320), who assigned the site to a single cultural complex, Santa Rosa-Swift Creek, in his monumental Gulf Coast synthesis, Bullen documented the multicomponent nature of the site. Bullen's excavations were principally in refuse deposits, where the earliest evidence of occupation was assigned to Late Deptford (not Swift Creek) culture (e.g., Milanich and Fairbanks 1980:75; Weisman 1995:5–9). Deptford Culture is now generally accepted as the context for the Yent Complex at the site (Milanich and Fairbanks 1980:85–6; Milanich 1994: 216). Three radiocarbon dates from Bullen's midden excavations, ranging between 30 B.C. ± 100 and A.D. 200 ± 130, pertain to the Deptford component and the associated Yent Complex (Weisman 1995:39).

Jerald Milanich, then Curator of Florida Archaeology at the Florida Museum of Natural History, provided eleven sherds from Crystal River for thin section analysis. These sherds come from various contexts that were tested by Bullen. These sherds are listed by type in Table 7.3, where their bulk and mineralogical compositions are recorded.

The initial review of these thin sections prior to point counting revealed at once that ten were sand tempered, while one clearly anomalous sherd was tempered with metagranite. To facilitate comparisons with thin sections from other sites on the Coastal Plain as well as with those from the Appalachians from where the granite-tempered vessel surely originated, bulk and mineralogical indices were used to record the compositions.

In Table 7.3 the ten stylistically local, sand-tempered vessels are tabulated as a group along with their two-sigma ranges. A single compositional value for each of two vessels falls outside of the 95 percent confidence limits: vessel 8-5 has a matrix value that is .2 percent beneath the relevant two-sigma range, while vessel 8-11 has a FMM value that is .4 percent higher than its relevant two-sigma limits. Such small departures from the two-sigma ranges are reasonably attributed to sampling error in both cases.

The ten local vessels present a relatively homogeneous compositional range with some distinctive properties. The most prominent feature of the group is the extraordinarily high incidence of monocrystalline quartz: a mean of 96.7 percent with no values below 92 percent. These values could certainly be considered diagnostic of northern, peninsular Florida.

Vessel 8-12 is definitely an import from the Appalachian mountain region (Figure 7.2). It is characterized by fragments of a granitic rock in a wide range of sizes, many polycrystalline but many monocrystalline as well (i.e., angular quartz or feldspar fragments). Distinguishing temper from natural sand proved to be impossible because all inclusions were mineral and rock grains derived from granitic rock, i.e., monocrystalline and polycrystalline quartz, granite (i.e., quartz + feldspars + amphiboles in various proportions), and monocrystalline feldspars and amphiboles. One distinctive feature is the heavy alteration that characterizes the feldspars. This vessel is similar compositionally to the Metagranite and Composite petrographic classes at the Leake site (cf. Tables 7.3 and 6.4), making Leake a leading candidate for the source of this vessel.

Interestingly, there is suggestive evidence that interaction between Crystal River and Leake may have been two way. From among the 37 vessels of the Monocrystalline Quartz Dominant class at the Leake site, one—Swift Creek

Table 7.3. Bulk and Mineralogical Composition of 11 Vessels from Crystal River (8Ci1)

Values in bold fall outside the relevant two-sigma ranges/95% confidence limits shown in shaded cells.

		Bulk Composition				*Mineralogy*		
Type	**Thin Section No.**	**% Matrix**	**% Silt**	**% Sand**	**Sand Size Index**	**% Monoc Quartz**	**% Polyc**	**% FMM**
Presumed Local Vessels								
Crooked River Complicated St	8-3	64	2	34	1.32	92	4	4
Deptford Simple St	8-4	75	1	24	2.00	100	0	0
Deptford Linear Check St	8-5	**52**	2	46	1.44	96	4	0
Deptford Simple St	8-6	62	1	37	2.21	99	1	0
Deptford Simple St	8-7	62	2	36	1.75	99	1	0
Deptford Simple St	8-8	70	2	28	1.83	93	5	2
Deptford Simple St	8-9	71	0	29	1.78	97	3	0
Plain	8-10	77	1	22	1.62	100	0	0
Plain	8-11	67	0	33	2.00	93	1	6
Unid. Punctated/Dentate St	8-13	74	3	23	1.67	98	2	0
Mean ±1 Std Deviation	[n=10]	67.4±7.6	1.4±1.0	31.2±7.5	1.76±.27	96.7±3.1	2.1±1.8	1.2±2.2
Two-Sigma Range	[n=10]	52.2-82.6	0-3.4	15.2-46.2	1.22-2.30	90.5-100	0-5.7	0-5.6
Probable Import								
Deptford (?) Simple St	8-12	63	**5**	32	**2.57**	**31**	**39**	**30**

Note: St=Stamped

7.2. Photomicrograph of Simple-Stamped Vessel 8-12 from Crystal River taken at 10X magnification under crossed polars. Large metagranite grain (with polycrystalline quartz, altered feldspar, and microcline) measures 2.235 mm. This vessel is probably derived from the Leake site.

Complicated Stamped vessel 9-270—had a Crystal River-like monocrystalline quartz percentage (93 percent). The other 36 vessels had 50–80 percent monocrystalline quartz (Stoltman 2010). This vessel stands out dramatically and deserves consideration as an intrusive from the Gulf region of peninsular Florida, perhaps even Crystal River. One drawback to the latter inference is that Swift Creek pottery is uncommon (though not completely absent; Bullen 1953:16–17) at Crystal River.

As a generalization, Middle Woodland ceramics from the Coastal Plain of Georgia and Florida are characterized predominantly by sand-sized inclusions of monocrystalline quartz—64 percent is the lowest incidence of monocrystalline quartz observed among the 26 thin-sectioned local vessels from Mandeville, Kolomoki, and Crystal River, and this pertained only to two vessels, both from Mandeville. The distinctiveness of the Mandeville ceramics may reasonably be attributed, at least partly, to its geographic location. As the northernmost of the three Coastal Plain sites in this study, the potters of Mandeville no doubt made use of alluvial sediments of the Chattahoochee River (which drains the Appalachian Mountains of north Georgia) that contained igneous and metamorphic minerals derived from those mountains.

The minimum amount of monocrystalline quartz that characterizes most Coastal Plain vessels is 75 percent, and this amount is accepted as a reasonable benchmark that must be met by an Ohio Valley vessel to be seriously considered as an import from the Coastal Plain. Whether these vessels were made from naturally sandy clays or were to some extent sand tempered is unknown at present. Either alternative is credible. In addition the relatively low incidence of FMM minerals (especially feldspars and amphiboles) is distinctive of Coastal Plain vessels. A maximum of 20 percent FMM minerals will be accepted as a reasonable upper limit for use in identifying vessels in the Ohio Valley as probable imports from the Coastal Plain.

8

Limestone-Tempered Pottery in Middle Woodland Contexts in the Southeast and the Ohio Valley Region

Fifty-three limestone-tempered vessels from Middle Woodland contexts were thin sectioned for this study. Twenty-four from the Ohio Valley region have already been considered—18 from Ohio (see Table 2.20), 5 from Mann (see Table 3.7), and one from Rutherford (see Chapter 4). Only paste indices were used in considering these vessels because of the incomparability of body indices for vessels with tempers as different as limestone, grit, and grog. As it turned out, with one notable exception, none of the Ohio and Mann limestone-tempered vessels had paste values that differed in significant ways from those of their local ceramics. From these data, it was inferred that most of these limestone-tempered vessels had been locally made within the Ohio Valley region where limestone bedrock is common. In this chapter a new dimension will be added to this discussion, namely, the search for possible external sources that could have exported limestone-tempered vessels into the Ohio Valley region. Before considering evidence of possible exchange of limestone-tempered pottery between the Ohio Valley and the Southeast, the compositional data for four Southeastern sites are first reviewed.

Thin sections of limestone-tempered pottery are available from four Southeastern sites, Icehouse Bottom (n=13), Tunacunnhee (n=7), Leake (n=7), and Pinson Mounds (n=2). Since temper could be reliably identified in these vessels (although in most cases this was done through the presence of polygonal voids, i.e., leached limestone, in the absence of any other temper), body as well as paste indices could be employed to characterize the compositions. Unfortunately, direct comparisons with the local pottery at each of these sites was not possible, because the fine textures necessitated the use of bulk composition rather than body and paste to characterize the composition of the local pottery. In theory direct compositional comparison with the limestone-tempered vessels of the Ohio region in terms of body and paste is possible because all 53 vessels share the same temper. This topic will be addressed at the end of this chapter, but first the compositional data for limestone-tempered pottery from each of the four Southeastern sites are presented in turn.

Limestone-Tempered Pottery from Icehouse Bottom

Icehouse Bottom is unique in many ways, not only because of the presence of imported vessels from Ohio but also for its intimate mixture of two Middle Woodland ceramic series—sand-tempered Connestee and limestone-tempered Candy Creek—that are rarely found together in eastern Tennessee (Chapman 1973:132; Chapman and Keel 1979:159). Because of the direct association of sherds from both ceramic series at Icehouse Bottom, 13 limestone-tempered sherds were included among those obtained for thin section analysis. Table 8.1 records the compositions of these 13 Candy Creek sherds, each representing a separate vessel.

The body and paste values are provided for each of the 13 vessels in Table 8.1, which are grouped by typological class: Wright Check Stamped (n=2); Bluff Creek Simple Stamped (n=9); and Candy Creek Cordmarked (n=2). Based upon these data, the two-sigma ranges (95 percent confidence limits) for each component of the compositional indices are recorded in the last line of Table 8.1. A comparison of each compositional value with each of the appropriate two-sigma ranges reveals that this sample of vessels is of generally uniform composition. Only two values (indicated in bold font) fall outside the relevant two-sigma ranges, thus identifying vessels 40-20 and 40-25 as possibly exotic. Closer examination of these values (i.e., an elevated sand size index for vessel 40-20 and a relatively high incidence of temper for vessel 40-25) most reasonably attributes these values to sampling error within the context of this relatively small sample of thin sections. In sum, it is reasonable to accept this sample of 13 thin sections as providing a reasonably reliable compositional signature for the Candy Creek ceramics at Icehouse Bottom.

The most notable compositional feature of the limestone-tempered pottery from Icehouse Bottom is the relatively high incidence (i.e., 8–10 percent) of sand in both paste and body indices (see Table 8.1 and Figure 8.1). In addition, the presence of minor amounts of clinozoicite in 11 of the 13 vessels, a ubiquitous inclusion in the local Connestee series at the site, may possibly serve as a marker mineral for the Candy Creek series at Icehouse Bottom in contrast to limestone-tempered ceramics from other sites.

Limestone-Tempered Pottery from Tunacunnhee and Leake in Northwest Georgia

Because of their close proximity, the limestone-tempered vessels from the two north Georgia sites are here considered together. There are seven such vessels from each site that were thin sectioned for this study. The body and paste indices for each of these vessels are recorded in Table 8.2.

Table 8.1. Body and Paste Values for 13 Limestone-Tempered Vessels from Icehouse Bottom

Values in bold exceed the relevant two-sigma ranges in the shaded cells.

		Body				*Paste*			
Type	**Thin Section No.**	**% Matrix**	**% Sand**	**% Temper**	**Temper Size Index**	**% Matrix**	**% Silt**	**% Sand**	**Sand Size Index**
Wright Check Stamped	40-19	78	4	18	2.97	92	3	5	1.25
Wright Check Stamped	40-20	79	7	14	3.05	86	6	8	**1.86**
Bluff Creek Simple St	40-23	74	10	16	3.71	81	7	12	1.10
Bluff Creek Simple St	40-24	84	8	8	3.42	84	7	9	1.24
Bluff Creek Simple St	40-25	67	2	**31**	3.87	95	1	4	1.50
Bluff Creek Simple St	40-26	71	5	24	3.75	86	7	7	1.56
Bluff Creek Simple St	40-27	79	8	13	3.12	82	9	9	1.47
Bluff Creek Simple St	40-28	63	13	24	3.79	81	2	17	1.32
Bluff Creek Simple St	40-29	71	16	13	3.23	74	8	18	1.59
Bluff Creek Simple St	40-30	80	10	10	3.73	81	8	11	1.15
Bluff Creek Simple St	40-31	76	9	15	3.31	84	5	11	1.18
Candy Creek Cordmarked	40-21	71	13	16	4.00	79	6	15	1.29
Candy Creek Cordmarked	40-22	78	5	17	4.00	90	4	6	1.27
Mean ±1 Std Deviation	n=13	74.7±5.9	8.5±4.0	16.8±6.3	3.54±.37	84.2±5.7	5.6±2.5	10.2±4.4	1.37±.22
Two-Sigma Ranges	n=13	62.9-86.5	0.5-16.5	4.2-29.1	2.80-4.28	72.8-96.6	0.6-10.6	1.4-19.0	1.00-1.81

Note: St=Stamped

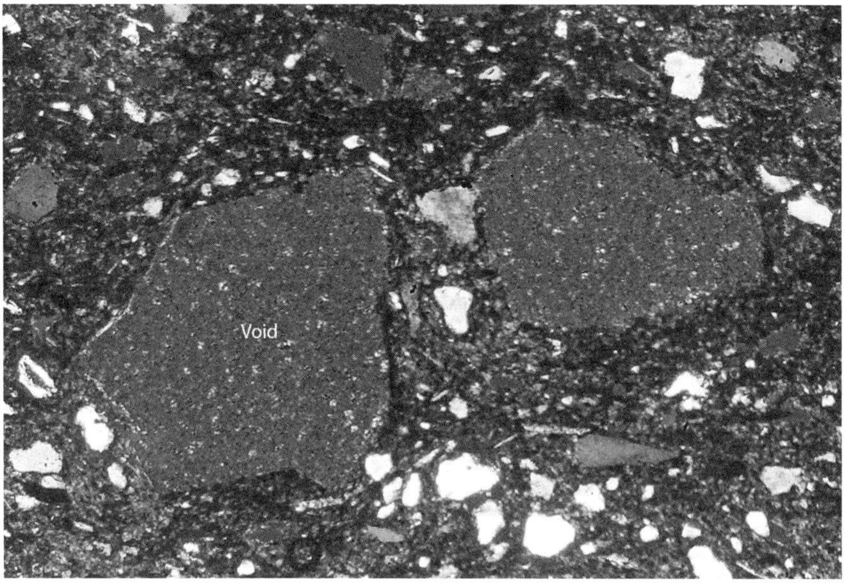

8.1. Photomicrograph of limestone-tempered Vessel 40-23 from Icehouse Bottom taken at 10X magnification under crossed polars. Note the two large voids from leached limestone, the relative abundance of both quartz sand and muscovite mica (prisms with high-order interference colors). Scale: large void on left=l.05 mm.

Checking each site first for possible compositional outliers by comparing individual values with the appropriate two-sigma ranges (95 percent confidence limits) recorded for each site in Table 8.2, it can be seen that two vessels, one from each site, have compositional values that fall outside of the relevant two-sigma ranges. Vessel 9-15 from Tunacunnhee has two conjoined values that fall outside of the relevant two-sigma ranges, with the elevated temper value of 37 percent being the primary determinant for the low matrix value. Since this value exceeds the relevant two-sigma range by less than one percentage point, the most reasonable explanation for such a high incidence of temper is sampling error or idiosyncrasies of local potters.

By contrast, Vessel 9-197 from Leake, which has unusually high sand percentages for both body and paste, presents a different interpretative problem. Such distinctively high sand values (9 percent and 11 percent respectively for body and paste; see Table 8.2) suggest that a distinctive clay source was used for this vessel. In contrast to the case of vessel 9-15, however, there is an external assemblage that has such high sand values, thus is a potential source for this vessel: Icehouse Bottom (see Table 8.1). In this case, then, it is reasonable to see vessel 9-197 from Leake as a possible import from Icehouse Bottom. As an addendum, it might be noted that clinozoicite is present in vessel 9-197.

Table 8.2. Body and Paste Values for Seven Limestone-Tempered Vessels Each from Tunacunnhee and Leake

Values in bold exceed the relevant two-sigma ranges shown in shaded cells.

		Body				*Paste*			
Type	**Thin Section No.**	**% Matrix**	**% Sand**	**% Temper**	**Temper Size Index**	**% Matrix**	**% Silt**	**% Sand**	**Sand Size Index**
TUNACUNNHEE									
Cordmarked	9-10	80	1	19	3.64	99	0	1	1.00
Cordmarked	9-15	**61**	2	**37**	4.08	91	5	4	1.00
Cordmarked	9-17	75	1	24	3.53	94	5	1	1.00
Cordmarked	9-18	77	1	22	3.60	91	8	1	1.00
Simple Stamped	9-16	75	3	21/1	2.70/4.00	95	1	4	1.00
Simple Stamped	9-20	80	1	16/3	3.29/3.75	89	9	2	1.00
Plain	9-11	76	1	23	3.79	96	2	2	1.00
Mean ±1 Std Deviation	n=7	74.9±6.6	1.4±0.8	23.7±6.2	3.52±.43	93.6±3.5	4.3±3.4	2.1±1.3	1.00±0
Two-Sigma Ranges	n=7	61.7–88.1	0–3.0	11.3–36.1	2.66–4.38	86.6–100	0–11.1	0–4.7	—
LEAKE									
Plain	9-193	83	0	17	2.97	96	4	0	—
Plain	9-194	80	1	19	3.57	92	7	1	4.00
Plain	9-195	80	2	18	3.54	91	7	2	1.33
Plain	9-197	75	**9**	16	3.41	84	5	**11**	1.07
Cordmarked	9-196	84	3	13	3.56	89	7	4	1.00
Fabricmarked	9-198	72	1	27	3.04	92	7	1	3.00
Fabricmarked	9-199	84	3	12/1	3.77/4.00	96	1	3	1.20
Mean ±1 Std Deviation	n=7	79.7±4.6	2.7±3.0	17.6±4.8	3.41±.30	91.4±4.2	5.4±2.3	3.2±3.7	1.93±1.26
Two-Sigma Ranges	n=7	70.5-88.9	0-8.7	8.0-27.2	2.81-4.01	83.0-99.8	0.8-10.0	0-10.6	1.00-4.45

As suspected from the close proximity of Tunacunnhee to Leake, the compositional values for the limestone-tempered pottery from the two sites are similar. There is, however, one exception—the sand size indices—that is tantalizing. The incidence of sand in the pastes at both sites is generally low—4 percent or less—with the exception of the suspected intrusive vessel (9-197) at Leake (see Table 8.2). It may indeed be significant that all seven vessels from Tunacunnhee have sand-size indices of 1.00. That is, all sand grains that occurred among the points that were counted were less than .25 mm in maximum diameter. Stated differently, no sand grains of medium or coarser size were included in the point counting for these vessels in contrast to Leake where six of the seven vessels had medium and/or coarse sand grains in the point counts. This is suggestive evidence that the clays used by Tunacunnhee potters were of finer texture than those used at Leake.

Pinson Mounds

One limestone-tempered vessel (40-65) and one possibly limestone-tempered vessel (40-67) are present among the thin-sectioned sherds from Pinson Mounds (Stoltman and Mainfort 2002: Fig. 2). The presence of limestone in both cases is inferred from the existence of voids presumed to represent leached limestone. The equivocation over vessel 40-67 has to do with the presence of biogenic structures (uncharred plant remains?) that partially fill some voids (three to be exact) (See Figure 6 in Stoltman and Mainfort 2002:15). The presence of these biogenic structures, even if only partially present in a few voids, interjects a note of uncertainty concerning the wisdom of accepting the remaining voids as derived from leached limestone.

The body and paste values for these two vessels are recorded in Table 8.3. Both sherds were among those recovered in the Duck's Nest sector that Mainfort (1986:35) regarded as "unequivocal examples" of non-local vessels. The rationale for this assessment is that Pinson Mounds is located in the West Tennessee Uplands physiographic province that is underlain by unconsolidated sediments—"mostly Cretaceous, Tertiary, and Quaternary gravel, sand, silt and clay"—with no bedrock exposures of any kind nearby (Floyd 1965:9 + geological map, p. 12, and Figure 15 opposite p. 72). Interestingly, both of these sherds were assigned to local Group 4 (INAA samples PMO067 and PMO071) in the neutron activation analysis of Pinson Mounds pottery (Mainfort et al. 1997:53).

Based upon their compositional differences, the two vessels are presumed to derive from different sources. Vessel 40-65 is unique in possessing a relatively high incidence of chert, which is here treated as temper since the bedrock source for most chert is limestone, and its paste is essentially devoid of

Table 8.3. Body and Paste Values for Two Limestone-Tempered Vessels from Pinson Mounds

		Body					Paste			
Type	**Thin Section #**	**% Matrix**	**% Sand**	**% Temper**	**Temper Size Index**	**Temper Type**	**% Matrix**	**% Silt**	**% Sand**	**Sand Size Index**
Untyped Corded	40-65	89	0	8	3.33	Limestone*	97	3	0	—
				3	3.43	Chert+				
Untyped Plain	40-67	82	8	9	3.04	Limestone*	86	5	9	1.00
				1	3.50	Plant				

* *Denotes leached limestone*

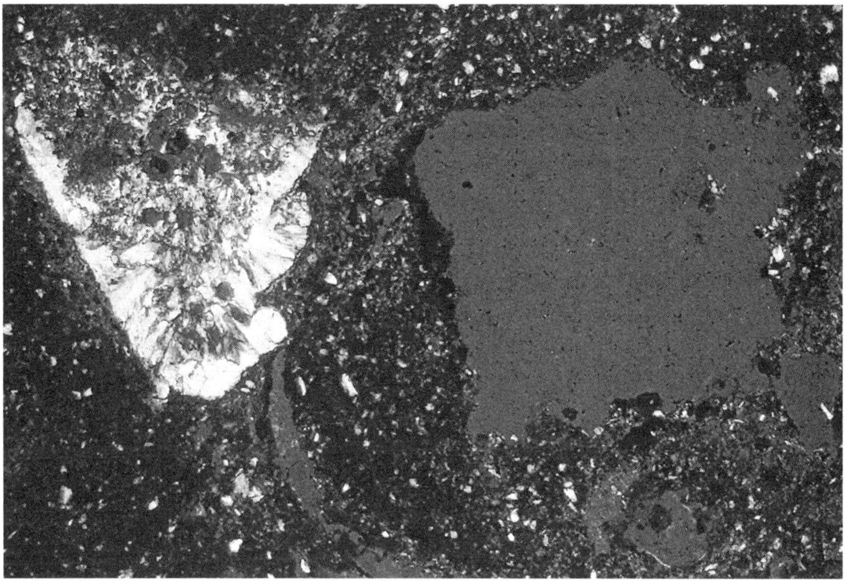

8.2. Photomicrograph of limestone-tempered Vessel 40-65 from Pinson Mounds taken at 10X magnification under crossed polars. Note the paucity of sand-size grains in the paste. Scale: chert grain on left=1.10 mm; large void on right [V] measures 1.325 mm.

sand (Figure 8.2). A Tennessee Valley source for this vessel is a reasonable inference.

There are several distinctive compositional properties of Vessel 40-67 that argue for its derivation from the Appalachian Mountain region: (1) muscovite ("white mica") is especially common, which is also true of the Candy Creek series at Icehouse Bottom; (2) clinozoicite, while not abundant, has multiple occurrences in this thin section; and (3) this vessel has unusually high frequencies of sand in both its body and paste, closely matching the Candy Creek series of Icehouse Bottom (cf. Tables 8.1 and 8.3). It is here suggested that Vessel 40-67 was indeed imported to the Pinson Mounds site and that the Appalachian Mountain region of eastern Tennessee is the likely general source.

Comparing Limestone-Tempered Pottery from the Southeast with the Ohio Valley Region

The four Southeastern sites that provided 29 limestone-tempered vessels for this study are relevant to the topic of Ohio Valley–Southeast interaction because of the unambiguous evidence of such interaction in other aspects of their assemblages. But, whether they provide an adequate sample for the task at hand is another matter. It must be admitted that the numbers of both sites and thin sections comprise samples of unknown representativeness of the Middle Woodland universe we are seeking to understand. But, these are the "cards we have been dealt," so we will do the best we can with what we have.

In considering the four Southeastern sites as possible exporters of limestone-tempered pottery to Ohio Valley sites, Pinson Mounds can be excluded because of the high probability that it was a consumer not a producer of such pottery. Its location on the Upper Gulf Coastal Plain affords it no direct access to limestone. In addition, both limestone-tempered vessels from the site that were thin sectioned (see Table 8.3) can reasonably be regarded as imports to the site, the INAA results notwithstanding.

The other three Southeastern sites, however, can surely be regarded as producers of limestone-tempered pottery, thus possible exporters as well. The compositional values for the limestone-tempered pottery from the Tunacunnhee and Leake sites (Table 8.2) so closely overlap those of the Ohio-Valley vessels (see Table 8.4) that any imports from these sites would be virtually undetectable. By contrast, Icehouse Bottom has a distinctive enough compositional signature that it may be used to identify possible exports. Its relatively high incidence of sand in both paste and body indices, coupled with the nearly ubiquitous presence of clinozoicite, albeit in trace amounts, provides reasonably firm evidence of the character of locally made limestone-tempered vessels at the site. As already discussed, there are possible cases of Candy Creek vessels having been recovered from two Southeastern sites, vessel 9-197 at Leake and vessel 40-67 at Pinson Mounds.

To facilitate comparisons between the Ohio Valley and Southeastern sites, Table 8.4 presents a compilation of body and paste values by site for each of the 24 Ohio Valley limestone-tempered vessels. It was evident from the outset that the limestone-tempered vessels from the Ohio Valley sites were broadly similar in their compositions. Most important, their paste values, with few exceptions, were similar to those of the local pottery at their respective sites, suggesting that most were manufactured locally. In the discussion to follow that these limestone-tempered vessels were made in the Ohio Valley region will be accepted as a null hypothesis, i.e., a proposition held to be true until contradicted by reliable evidence (Bhattacharyya and Johnson 1977:167–168).

In considering possible interaction between the Ohio Valley sites and the Southeastern sites, three basic principles will guide the discussion: First, in line with the null hypothesis, those Ohio-Valley vessels with compositional values that all fall within the 95 percent confidence limits of the regional group recorded at the bottom of Table 8.4 are accepted as locally made. Second, the search for possibly imported vessels will be limited to those that have compositional values that fall outside of the relevant two-sigma ranges. And third, only vessels with compositional values that fall beyond the 95 percent confidence limits of the Ohio Valley group *and* conform to those of an identifiable external source will be proffered as "probable imports."

From Table 8.4 it can be seen that only four of 24 vessels—no. 33-99 from Mound City, no. 33-124 from Seip, no. 33-275 from Turner, and no. 77 from Mann—have compositional values that fall beyond the 95 percent confidence limits derived from the group means. One of these vessels, McGraw Cordmarked no. 33-99, has an elevated silt value suggestive of the use of a distinctively silty clay. As discussed in Chapter 3, such high-silt clays were sometimes used in Chillicothe-region ceramics so a local origin is likely. Another of these vessels—no. 33-124 from Seip—has an elevated temper value in its body that could be a reflection of idiosyncratic pottery-making practices by a potter or possibly sampling error (e.g., a thin section derived from an atypical part of a vessel). In any case there are currently no known external sources characterized by compositional properties observed in vessels no. 33-99 and no. 33-124 that could be their source. The third vessel in Table 8.4 with anomalous compositional values, coarse simple-stamped vessel no. 77 from Mann, has a sand size index that can be attributed to sampling error due to an unusually large sand grain occurring in a thin section with relatively few sand grains. Its small temper size index of 2.23, however, is a somewhat different matter, for its limestone inclusions contain apatite, something seen in no other vessel. As with the foregoing two vessels, no known possible external source for this vessel is known. Finally we come to McGraw Plain vessel no. 33-275 from the Turner site, which is an excellent candidate for an intrusive vessel. Clearly, the clay from which it was made is distinctive within the context of the present sample of limestone-tempered vessels as is evidenced by the notably elevated sand values for both body and paste (see Table 8.4). In this sense it possesses a distinctive paste like that characterizing the Candy Creek series at Icehouse Bottom (cf. Tables 8.1 and 8.4). Moreover, at least one clinozoicite grain occurs in this Turner vessel, something that characterizes none of the other Ohio Valley vessels.

In this discussion of possible transport or exchange of limestone-tempered pottery in Middle Woodland contexts, Icehouse Bottom has emerged as the most active participant. The extensively overlapping compositional values of the limestone-tempered pottery among the Tunacunnhee, Leake, and Ohio Valley sites mitigates against any realistic discussion of interaction among these sites. It is more parsimonious to accept the null hypothesis of the local manufacture of Ohio Valley pottery than to postulate exchange. As for Ice-

Table 8.4. Paste and Body Values for All [n=24] Limestone-Tempered Vessels from Ohio Valley Sites

Values in bold exceed the relevant ± two-sigma ranges shown in shaded cells.

Paste

Site/Type	Thin Section No.	% Matrix	% Silt	% Sand	Sand Size Index
Mound City	**[n=2]**				
Turner Check Stamped	33-79	97	2	1	1.00
McGraw Cordmarked	33-99	82	**17**	1	1.00
Liberty	**[n=4]**				
McGraw CM (RB-Md 2)	33-240	82	9	9	1.71
McGraw CM (RB-Md 2)	33-249	85	10	5	1.89
Turner S St. A (Harness)	33-131	94	1	5	1.25
Turner S St. A (Harness)	33-144	92	1	7	1.22
Seip	**[n=3]**				
McGraw Plain	33-123	95	4	1	1.00
McGraw Plain	33-124	99	1	0	—
Turner Check Stamped	33-198	96	4	0	—
Marietta	**[n=1]**				
McGraw Cordmarked	33-293	88	8	4	2.00
Turner	**[n=8]**				
Hopewell Rim	33-276	97	3	0	—
Chillicothe Rocker Stamped	33-278	90	9	1	1.50
McGraw Cordmarked	33-257	100	0	0	—
McGraw Cordmarked	33-258	89	6	5	1.67
McGraw Cordmarked	33-259	93	6	1	2.50
McGraw Plain	33-270	87	7	6	1.27
McGraw Plain	33-274	94	6	0	—
McGraw Plain	33-275	**71**	10	**19**	1.91
Mann	**[n=5]**				
Cordmarked	25	93	5	2	1.00
Plain	49	82	9	9	1.33
Plain	50	95	5	0	—
Plain	54	97	3	0	—
Coarse Simple Stamped	77	94	4	2	**3.00**
Rutherford	**[n=1]**				
Corddmarked	11-243	90	9	1	1.00
Mean ±1 Std Deviation	**[n=24]**	90.9±6.8	5.8±3.9	3.3±4.4	1.52 ±.59
Two-Sigma Range/Paste	**[n=24]**	77.3-100	0-13.6	0-12.1	1.00-2.70

Note: S St.=Simple Stamped; CM=Cordmarked

Body

Site/Type	Thin Section #	% Matrix	% Sand	% Temper	Temper Size Index
Mound City	**[n=2]**				
Turner Check Stamped	33-79	82	1	17	3.04*
McGraw Cordmarked	33-99	84	1	15	4.10
Liberty	**[n=4]**				
McGraw CM (RB-Md 2)	33-240	89	8	3	4.50
McGraw CM (RB-Md 2)	33-249	93	5	2	3.33*
Turner S St. A (E. Harness)	33-131	74	4	22	3.83
Turner SSt. A (E. Harness)	33-144	77	6	17	3.82
Seip	**[n=3]**				
McGraw Plain	33-123	74	1	25	4.00
McGraw Plain	33-124	**59**	0	**41**	4.78
Turner Check Stamped	33-198	77	0	23	2.53
Marietta	**[n=1]**				
McGraw Cordmarked	33-293	83	4	13	3.62
Turner	**[n=8]**				
Hopewell rim	33-276	85	0	15	3.40
Chillicothe Rocker Stamped	33-278	84	1	15	3.80
McGraw Cordmarked	33-257	72	0	28	2.71
McGraw Cordmarked	33-258	75	4	21	3.52
McGraw Cordmarked	33-259	84	1	15	3.73
McGraw Plain	33-270	74	5	21	3.83
McGraw Plain	33-274	89	0	11	3.00
McGraw Plain	33-275	70	**17**	13	3.80
Mann	**[n=5]**				
Cordmarked	25	76	1	23	3.62
Plain	49	82	8	10	4.12
Plain	50	97	0	3	3.40
Plain	54	80	0	20	4.31
Coarse Simple Stamped	77	74	2	24	**2.23**
Rutherford	**[n=1]**				
Cordmarked	11-243	96	1	3	3.75
Mean ±1 Std Deviation	**[n=24]**	80.4±8.8	2.9±3.9	16.7±9.1	3.62±.60
Two-Sigma Range/Body	**[n=24]**	62.8-98.0	0-10.7	0-34.9	2.42-4.82

**Denotes leached limestone*

house Bottom, the combination of relatively high sand values (i.e., >9 percent on average) in both paste and body combined with the presence of the metamorphic mineral, clinozoicite, (which can only come from the crystalline Precambrian rocks of the Blue Ridge province) provides a reasonably reliable compositional signature for the Candy Creek series at that site. Using this compositional signature, it was possible to identify vessels from three different sites, Leake, Pinson Mounds, and Turner, as probable recipients of limestone-tempered pottery vessels from Icehouse Bottom or its vicinity.

9
Pinson Mounds Revisited

A reconsideration of the Pinson Mounds thin section data previously published (Stoltman and Mainfort 2002) is presented here in order to clarify some issues, to add mineralogical compositions not included in the earlier report, and to view the site in the larger context provided by the newer data in this volume. A certain amount of notoriety has accrued to the site because of a unique set of circumstances, at least at that time, involving the neutron activation analysis (NAA) of 128 pottery sherds from Pinson Mounds that were all considered "probably of local origin" (Mainfort et al. 1997:62) followed by a reevaluation of this conclusion based on the petrographic analysis of 32 sherds from the site of which 22 were among those previously subjected to NAA (Stoltman and Mainfort 2002). The relative merits and limitations of NAA and petrography were discussed in the 2002 paper and will not be revisited here. My goal in this chapter is to amplify the earlier findings from petrography to demonstrate that, contrary to the conclusions of the neutron activation analysis that "all analyzed pottery from Pinson Mounds was produced locally" (Mainfort et al. 1997:43), several vessels were, indeed, imported to the site. By documenting that pottery was imported to Pinson Mounds from several different sources, the goal is to emphasize the importance of the site as a major regional center of cultural interaction, contrary to the implications of the NAA findings.

In the original petrographic analysis only bulk compositional data were used to characterize the thin sections. With the addition of the mineralogical indices, important new perspectives on these data are now possible. In this chapter, in contrast to the original two studies, the focus will be entirely upon Pinson Mounds, so the data from nearby sites will not be considered.

Table 9.1 presents the bulk and mineralogical compositions for 14 vessels identified as local by Mainfort because they manifest the two salient properties that characterize the site's pottery: sand temper and distinctive types of surface treatment (i.e., cordmarking, fabric impressing, and plain), with little

Table 9.1. Bulk and Mineralogical Composition of 14 Stylistically Local, Sand-Tempered Vessels from Pinson Mounds

Values in bold fall beyond the relevant two-sigma ranges.

Type	Thin Section No.	*Bulk Composition* % Matrix	% Silt	% Sand	Sand Size Index	*Mineralogy* % Monoc Quartz	% Polyc	% FMM
Local Vessels								
Furrs Cordmarked	40-46	69	9	22	1.63	85	**15**	0
Furrs Cordmarked	40-50	82	7	11	1.00	86	7	7
Furrs Cordmarked	40-51	68	6	26	1.66	89	8	3
Furrs Cordmarked	40-52	78	2	20	1.58	84	10	6
Furrs Cordmarked	40-64○	67	9	24	1.43	86	12	2
Furrs Cordmarked	40-88■	84	2	14	1.78	84	8	8
Furrs Cordmarked	40-89■	81	2	17	1.85	84	12	4
Furrs Cordmarked	40-90■	71	6	23	1.05	82	9	9
Baldwin Plain	40-45	70	8	22	1.63	85	12	3
Baldwin Plain	40-48	72	5	23	1.49	93	7	0
Saltillo Fabric Imp.	40-91○	72	11	17	1.88	**95**	5	0
Mean ±1 Std Deviation	**[n=11]**	74.0±6.1	6.1±3.1	19.9±4.6	1.54±.29	86.6±4.1	9.6±2.9	3.8±3.3
Two-Sigma Ranges	**[n=11]**	61.8-86.2	0-12.3	10.7-29.1	1.00-2.12	78.4-94.8	3.8-15.4	0-10.4
Suspected Nonlocal								
Furrs Cordmarked	40-43↗	73	7	20	1.09	80	7	**13**
Furrs Cordmarked	40-44↗	80	7	13	1.16	**68**	10	**22**
Saltillo Fabric Imp.	40-47↗	**60**	3	**37**	1.83	79	8	**13**

Note: ↗ =*Vessels with micaceous pastes and exotic minerals; considered probably nonlocal.* ○=*Vessels "Unassigned" by neutron activation (Mainfort et al. 1997)* ■=*Vessels identified as local by neutron activation (Mainfort et al. 1997)* *Imp.=Impressed*

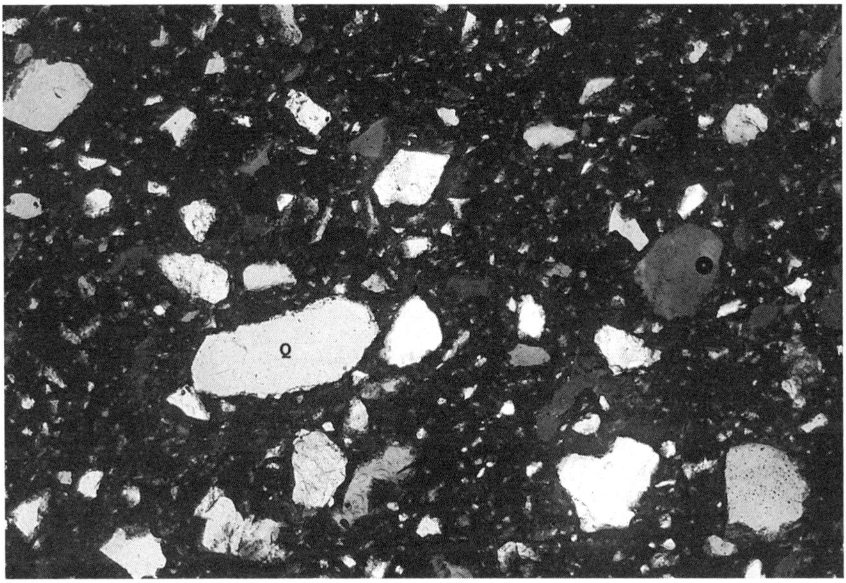

9.1. Photomicrograph of Baldwin Plain Vessel 40-48 from Pinson Mounds taken at 10X magnification under crossed polars. Note the predominance of monocrystalline quartz and the absence of mica and most other FMM minerals. Scale: largest quartz grain near center [Q]= .55mm.

or no decoration (see also Mainfort 2013:235). Figure 9.1 shows the characteristic compositional properties of the local vessels at Pinson Mounds, i.e., a predominance of monocrystalline quartz and a paucity of FMM minerals. With the mineralogical indices now added for consideration, three of the 14 "local" vessels (rather than just one, Vessel 40-47) appear to be of nonlocal composition (Vessels 40-43, 40-44, and 40-47 in Table 9.1). What convincingly distinguishes these three vessels is their high mica content (both muscovite and biotite) along with some feldspars, and the presence of the distinctive epidote-family minerals, clinozoicite and epidote (Figure 9.2). While monocrystalline quartz dominates the mineralogical inventory of these three vessels, the elevated content of FMM minerals, whose original source must be the Appalachian Mountains, indicates that these vessels were not made from local clays. Vessel 40-47 is also distinctive in possessing an unusually high incidence of sand, nearly double the value of the mean for the local vessels. It seems likely that these three vessels were imports to Pinson Mounds, something that was only suspected for Vessel 40-47 in the original report (Stoltman and Mainfort 2002:7). With reference to Table 6.4, which contains the two-sigma ranges for the Monocrystalline-Quartz-Dominant petrographic

9.2. Photomicrograph of cordmarked vessel 40-43 from Pinson Mounds taken at 25X magnification under crossed polars. Despite the much smaller area covered by figure 9.2. as contrasted with figure 9.1, note the presence of several prismatically shaped grains of mica [M] that are characterized by parallel extinction, plagioclase feldspar [P] (twinned grain at upper right), and the small grain of clinozoicite [Cz] (just to right of center with high relief), all minerals of Appalachian origins. Scale: twinned plagioclase [P]=.175 mm; clinozoicite [Cz]=.05mm.

class at Leake, it can be seen that all compositional values except one for these three Pinson Mounds vessels fall fully within the two-sigma ranges for this Leake petrographic class. The one exception (the 1.09 sand-size index for Vessel 40-43, which falls just below the 1.10 lower two-sigma limit for Leake sand-size indices) is attributable to sampling error, making the Leake site a realistic source from which these vessels originated.

Let us turn now to two groups of vessels suspected of being of nonlocal origin. The first consists of sand-tempered vessels that have nonlocal types of decoration as recorded in Table 4 of the original report (Stoltman and Mainfort 2002:11). Table 9.2 records the compositional data for these same 11 vessels, this time with the addition of the mineralogy indices, and compares them to the two-sigma/95 percent confidence limits for the eleven local vessels as recorded in Table 9.1. In the original report we accepted the bulk compositional data at face value to suggest that two of these vessels (nos. 40-

Table 9.2. Bulk and Mineralogical Composition of 11 Stylistically Nonlocal, Sand-Tempered Vessels Compared to the Two-Sigma Range for Local Vessels from Pinson Mounds

Values in bold fall beyond the relevant two-sigma limits shown in shaded cells.

		Bulk Composition				*Mineralogy*		
Type	**Thin Section #**	**% Matrix**	**% Silt**	**% Sand**	**Sand Size Index**	**% Monoc Quartz**	**% Polyc**	**% FMM**
Swift Creek Complicated Stamped	40-56	65	6	29	1.44	82	11	7
Swift Creek Complicated Stamped	40-57	65	11	24	1.42	87	8	5
Swift Creek Complicated Stamped	40-58	65	9	26	1.55	84	**16**	0
Swift Creek Complicated Stamped	40-59	**58**	**13**	29	1.46	84	10	6
Untyped Simple Stamped	40-49	**61**	7	**32**	1.45	83	13	4
Burnished	40-60	65	7	28	1.45	85	12	3
McLeod Simple Stamped	40-62	70	6	24	1.52	82	**18**	0
McLeod Simple Stamped	40-63	73	5	22	1.23	93	7	0
Marksville Stamped	40-68	70	7	23	1.93	87	**3**	10
Marksville Stamped	40-71	**59**	**13**	28	1.04	78	**20**	2
Untyped Check Stamped	40-73	74	7	19	1.69	93	7	0
Pinson Mds Two-Sigma Ranges	**[n=11]**	61.8-86.2	0-12.3	10.7-29.1	1.00-2.12	78.4-94.8	3.8-15.4	0-10.4

Note: All vessels except 40-49 (which was not neutron activated) were identified as local by neutron activation analysis (Mainfort et al. 1997).

49 and 40-71) "appear to be nonlocal" (Stoltman and Mainfort 2002:11). In addition, while not explicitly citing the complicated stamped sherds as of nonlocal origin, we opined that such an origin "should not be completely discounted" (Stoltman and Mainfort 2002:15). The data in Table 9.2 shed considerable new light on this issue, which has been clouded with misrepresentations and misunderstanding (e.g., Neff et al. 2006:70).

As has been done throughout this volume, individual compositional values that fall outside of the relevant two-sigma limits (in this case, for the 11 local vessels recorded in Table 9.1) are shown in bold font. Six of these 11 vessels have such values (Vessels 40-58, 40-59, 40-49, 40-62, 40-68, and 40-71), raising the possibility of their external origins, while five have compositional values that all fall within the relevant two-sigma ranges (nos. 40-56, 40-57, 40-60, 40-63, and 40-73) suggesting their local origins. Close examination of the anomalous values reveals that most are minor, easily accounted for by sampling error in the context of small samples. This is the case for Vessels 40-49, 40-58, 40-59, 40-62, and 40-68, all with minor (i.e., <4 percent) departures from their relevant two-sigma values. There are certainly no known sites that can be cited as possible sources for these vessels. One vessel (no. 40-71), however, has what seems to be a significantly elevated polycrystalline value that is anomalous. Most of this is due to multiple grains of chert and somewhat coarser-grained polycrystalline quartz, so the unusually high polycrystalline count in Vessel 40-71 is not attributable to exotic (i.e., nonlocal) rocks. Based upon these considerations of the new data in Table 9.2, all 11 vessels are accepted as local products. In this regard I accept as valid the criticism raised by Neff et al. 2006:70 that the anomalous bulk compositional values for three vessels (these are the same in Table 4 in Stoltman and Mainfort 2002:11 and Table 9.2 in this chapter) as "slim evidence on which to base the inference of importation." Unfortunately, apparently Neff et al. 2006:70 failed to take note of the rest of the paper, completely neglecting to consider the more robust data originally presented in Table 3 in Stoltman and Mainfort 2002:10 (Table 9.3).

The second group of vessels suspected of being of nonlocal derivation are those tempered with materials other than sand. The bulk and mineralogical compositions of the 7 vessels in this group are recorded in Table 9.3. The bulk (not body) compositions for this group of vessels were reported in Table 3 of the original report (Stoltman and Mainfort 2005:10).

With the mineralogical data now included, at least five of the seven appear as strong candidates for intrusive vessels. Most prominent is red-filmed, grog-tempered Vessel 40-66, which stands out in every way as anomalous at Pinson Mounds (see Figure 3.4). Equally significantly, its composition finds

Table 9.3. Bulk and Mineralogical Composition of Seven Non-Sand-Tempered Vessels from Pinson Mounds

Values in bold fall beyond the relevant two-sigma limits for Local Vessels from Pinson Mounds shown at the bottom of the table

		Bulk Composition				***Mineralogy***		
Type/Temper	**Thin Section No.**	**% Matrix**	**% Silt**	**% Sand**	**Sand Size Index**	**% Monoc Quartz**	**% Polyc**	**% FMM**
Corded/Quartzite	40-61↗	**60**	**16**	24	**2.77**	**28**	**64**	8
Corded/Limestone	40-65■↗	86	3	11	**3.36**	**0**	**100★**	**0**
Red Filmed/Grog	40-66■↗	72	11	17	**2.65**	**30**	**70**	0
Plain/Limestone + Plant	40-67■↗	77	5	18	1.98	**37**	**60★**	3
Incised/Bone	40-69■+	69	**14**	17	1.51	**70**	**28**	2
Check Stamped/Quartzite	40-70■↗	**58**	6	**36**	**2.28**	**70**	**22**	8
Check Stamped/Grog	40-72■+	64	7	29	**2.36**	81	13	6
Pinson Mounds Local Two-Sigma Ranges	[n=11]	61.8-86.2	0-12.3	10.7-29.1	1.00-2.12	78.4-94.8	3.8-15.4	0-10.4

Note:

■=*Vessels identified as local by neutron activation (Mainfort et al. 1997)*

↗=*Vessels considered of nonlocal origin.*

+=*Vessels accepted as probably local.*

★=*Some due to leached limestone.*

a match at an external source, the Mann site. As discussed in Chapter 4, this vessel is regarded as one of several that were manufactured at the Mann site but recovered at other Southeastern sites.

Two of the vessels in Table 9.3 (nos. 40-65 and 40-67) are regarded as limestone tempered, one with high probability (40-65), the other (40-67) with some reservations. With the nearest source for limestone being the Tennessee River valley some 50 km to the east (Floyd 1965:Figure 15), the nonlocal production of these vessels is still acceptable as "a certainty" (Mainfort et al. 1997:45; See also Mainfort 1986:35). Vessel 40-65 could not be more un-Pinson-like with its nearly quartz-free paste (Table 9.3; see Figure 8.2). Vessel 40-67, which was previously discussed in Chapter 8, in contrast has several mineral inclusions that point ultimately toward an Appalachian source no matter whether some of the voids are attributable to leached limestone or all are accepted as derived from degraded plant material.

Two of the vessels in Table 9.3 (nos. 40-61 and 40-70) are tempered with quartzite, a metamorphic rock that is unambiguously exotic to the Pinson Mounds region (Figure 9.3). The compositional values suggest the two vessels do not share a common source. Neither is close enough in composition to the quartzite-tempered vessels from Tunacunnhee (cf. Table 6.2) or Leake (cf. Table 6.4) to implicate those sites as their source. Thus, these two vessels are accepted as nonlocal but with the source presently unknown. That Vessel 40-70 is also stylistically exotic (i.e., check stamped) is consistent with the compositional evidence of its nonlocal origin.

The sixth vessel in this group, no. 40-69, is unique in having incised decoration and bone inclusions, so seems a reasonable candidate for "possibly nonlocal" status. The bone occurs in small enough amounts (2 percent of bulk composition) that it could be incidental. In that case this vessel could have been made locally, considering that bone, unlike quartzite for example, was surely locally available at Pinson Mounds. The nonlocal status of this vessel remains a tenuous possibility.

The final vessel in Table 9.3 (no. 40-72) has local compositional values for all indices except sand size, which is due to several coarse grains of monocrystalline quartz plus two coarse grog grains. The grog is best regarded as incidental, meaning it is too sparse to allow the distinction to be made between accidental inclusions and intentional temper. With the exception of the unusual abundance of coarse quartz in Vessel 40-72 (which can easily be attributed to the thin section being a biased sample of the vessel from which it was derived), when viewed in the context of the quartz-rich character of Pinson Mounds pottery generally, a local manufacture for this vessel is the simplest and most acceptable view.

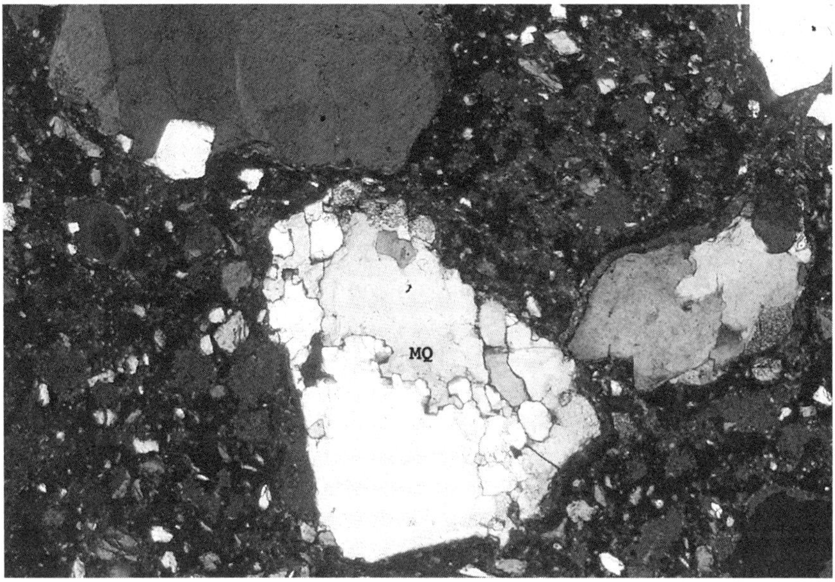

9.3. Photomicrograph of quartzite temper in check-stamped Vessel 40-70 from Pinson Mounds taken at 10X magnification under crossed polars. Scale: large central metaquartzite grain [MQ]= 1.10mm. Note sutured boundaries and undulose extinction, classic properties of metamorphic rocks.

Conclusions

One of the more interesting outcomes of this reconsideration of the Pinson Mounds thin sections concerns those vessels that were suspected to be nonlocal products because of their exotic forms of decoration (i.e., the 11 vessels in Table 9.2). As discussed above, a reconsideration of the compositional data for these vessels has led to a revised opinion of their origins; namely, that all 11 were probably made locally. This includes Vessels 40-49 and 40-71 as well as the four complicated stamped sherds.

This finding with respect to the complicated stamped sherds is consistent with a growing body of evidence to the effect that carved paddles (or the idea of making carved paddles) circulated widely among Middle Woodland peoples in the Southeast unassociated with the movement of pottery containers. The movement of paddles rather than pots has been documented among Swift Creek sites on the Georgia coast within the known range of Swift Creek culture (See Stoltman and Snow 1998). However, the Pinson Mounds findings are consistent with recent data from the Mann site (Ruby and Shriner 2004 and Chapter 3, this volume) in demonstrating this point even more dramatically by confirming that Swift Creek-like vessels found well beyond the normal range of such pottery were not imported but were made of local materials.

In addition to these findings the current reevaluation has provided a strong evidential basis for regarding at least three vessels from Table 9.1 and five of the seven vessels in Table 9.3 as probable intrusives to Pinson Mounds. For whatever reason, the bulk compositional data presented in Table 4 of the original report (Stoltman and Mainfort 2002:11; see also Table 9.2 in this chapter) became the main target of criticism (e.g., Neff et al. 2006:70), thus deflecting the debate from the more robust data. The main issue being challenged here is not the relative merits of NAA for ceramic analysis (that can be, and has been, debated elsewhere), but the implications that derive from the view that Pinson Mounds made all of its own pottery and that none was imported from elsewhere. Such a view of self-containment and isolation for Pinson Mounds is simply unacceptable.

10

The Sources of the "Possibly Nonlocal" Vessels Recovered from the Ohio Valley Sites

Compositional Profiles of Ohio Versus Southeastern Ceramics

In attempting to distinguish Southeastern imports from local Ohio Hopewell ceramics, two salient properties stand out; namely, Southeastern vessels are generally endowed with (1) a greater abundance of sand-size inclusions that are (2) finer in size. Tables 10.1 and 10.2 provide quantitative data to give substance to these observations.

These tables record the mean size indices for all counted sand-sized grains in each vessel (i.e., both temper *and* natural sands) and the mean sand densities (i.e., the total number of sand grains counted in each thin section standardized to 100 points counted at 1-mm intervals) for all local vessels from seven Ohio sites (Table 10.1) and eight Southeastern sites (Table 10.2). As grain-count rather than volumetric measures, these indices differ from those used in the analysis of the Ohio Hopewell vessels in two ways: (1) the size index includes *all* sand-sized grains counted in each thin section, not separate values for temper and natural sand; and (2) all sand-sized grains, even those receiving multiple counts, are recorded as single grains. Adopting these procedures, direct comparisons between the Ohio and Southeastern thin sections are possible that provide estimates of (a) the average size of all counted sand-sized grains in a thin section and (b) the total number of sand-sized grains within each thin section per 100 counts at 1-mm intervals.

Focusing first on the mean sand-size indices recorded in Tables 10.1 and 10.2, it can be seen at a glance that Ohio Hopewell thin sections generally have values above 3.00 whereas thin sections from Southeastern Middle Woodland sites generally have values between 1.00 and 2.00. There are some exceptions to these generalizations, which bear comment. The only mean size indices for vessels from Ohio Hopewell sites that are under 3.00 pertain to Hopewell-series vessels, with the sole exception being the Turner-series vessels at the Turner site (see Table 10.1). This is simply a further expression of the pattern noted earlier with respect to temper sizes and amounts, namely, that

Table 10.1. Mean Sand Size Indices and Sand Density Values [Grains/100 1-mm Counts] for Seven Major Middle Woodland Sites with Grit-Tempered Ceramics in Ohio

Site/Ceramic Series	N	Sand Size Index	Sand Density
Hopewell [n=13]			
Scioto Series	8	3.16±.16	18.5±5.3
Hopewell Series	3	2.30±.52	19.1±7.2
Turner Series	2	3.38±.43	11.4±0.9
Liberty [n=38]			
Scioto Series	18	3.14±.32	20.2±4.2
Hopewell Series	13	2.94±.26	17.0±4.4
Turner Series	7	3.03±.16	16.6±2.5
Seip [n=39]			
Scioto Series	16	3.40±.34	17.6±4.6
Hopewell Series	14	3.12±.33	15.6±3.0
Turner Series	9	3.20±.30	18.5±6.6
Mound City [n=26]			
Scioto Series	19	3.14±.38	16.8±5.0
Hopewell Series	6	2.59±.26	16.4±2.6
Turner Series	1	3.40	29.9
Turner [n=13]			
Scioto Series	7	3.15±.41	17.9±2.0
Hopewell Series	4	3.66±.23	12.9±2.0
Turner Series	2	2.81±.37	22.6±8.8
McGraw [n=6]			
Scioto Series	5	3.00±.62	10.9±8.3
Hopewell Series	1	2.96	12.0
Marietta [n=9]			
Scioto Series	9	3.47±.41	8.3±2.6
Totals	144		

Hopewell-series vessels are generally finer grained than vessels of the Scioto and Turner series. Among the Southeast vessels, there are but three exceptions to the below-2.00 "rule" for mean size indices (see Table 10.2). Two of these exceptions pertain to vessels from the two north Georgia sites where metaquartzite, an extremely hard, difficult-to-crush metamorphic rock, is the predominant inclusion. The third exception pertains to the Hopewell series at Garden Creek, which has a mean sand-size index of 2.14 (see Table 10.2). It is interesting to speculate that the craftspersons making these vessels were familiar with traditional pottery-making practices in both North Carolina

Table 10.2. Mean Sand Size Indices and Sand Density Values [Grains/100 1-mm Counts] for Eight Major Middle Woodland Sites in the Southeast

Site/Ceramic Series	N	Sand Size Index	Sand Density
Garden Creek			
Connestee Series	12	1.93±.16	33.1±5.2
Hopewell Series	8	2.14±.24	31.7±3.9
Turner Simple Stamped	2	1.12±.02	29.6±4.1
Icehouse Bottom			
Connestee Series	9	1.36±.16	24.8±4.4
Tunacunnhee			
Fissured Granite (Grit 1)	8	1.96±.12	33.3±2.9
Metaquartzite (Grit 2)	4	2.80±.20	24.4±2.4
Leake (Petrographic Classes)			
Monocrystalline Quartz Dom	37	1.62±.26	25.6±7.5
Metagranite	23	1.94±.24	28.1±9.9
Amphibole Rich	14	1.56±.17	25.4±6.7
Metaquartzite	4	2.01±.13	28.5±7.3
Composite	47	1.61±.24	22.1±6.8
Opaque Rich	1	1.70	15.0
Mandeville			
Local Vessels	10	1.55±.37	27.7±8.6
Kolomoki			
Swift Creek Series	6	1.56±.34	31.5±7.2
Crystal River			
Deptford Series	10	1.76±.27	31.2±7.5
Pinson Mounds			
Local Vessels	14	1.50±.31	20.6±6.6
Grand Total	209		

and Ohio and that these stylistically Ohio-like vessels made from local North Carolina materials reflect this interaction.

Focusing next on the sand density indices recorded in Tables 10.1 and 10.2, a clear pattern is once again evident; namely, the preponderance of Ohio vessels have mean sand density values less than 20 per 100 1-mm counts, while the majority of Southeastern vessels have sand density values greater than 20 per 100 1-mm counts. The exceptions are minor and probably attributable to sampling error associated with small samples.

From these data, what can be referred to as compositional profiles may be formulated to distinguish grit from sand/fine grit-tempered, Middle Woodland pottery from Ohio and the Southeast. On the one hand vessels that have bulk sand-size indices greater than 3.00 and sand density indices of less than 20 sand grains/100 mm have Ohio profiles. On the other hand, vessels with bulk sand-size indices between 1.00 and 2.00 and sand density indices greater than 20 sand grains/100 mm have a Southeastern profile.

These differences are significant and suggest the existence of two distinctly different traditions of pottery making. Southeastern ceramics are, by and large, extremely fine grained and are generally referred to as sand tempered. This is certainly true of the Coastal Plain and the Pinson Mounds region where monocrystalline quartz is the predominant mineral inclusion in pottery vessels. Since potters in these regions lack direct access to igneous and metamorphic rocks, their pottery will normally be made from alluvial clays that contain an abundance of redeposited sand grains, mostly quartz.

It is understandable that in such settings where sand grains have been transported and redeposited, pottery vessels would have an abundance of relatively fine-grained inclusions. But why, in the mountainous landscape of Appalachia, do the native ceramics have similar compositional profiles (see Table 10.2)? Answers to this question may differ if it were known whether these Southeastern ceramics were truly tempered with finely crushed grit or if the mineral inclusions in these vessels were derived directly from alluvial sediments. Either way, however, it appears that Southeastern potters carefully prepared and/or selected the materials used for pottery manufacture and that those choices reflected culturally conditioned practices that maximized the use of an abundance of fine-grained inclusions. By contrast, Hopewellian potters in Ohio selected and prepared coarse-grained crystalline rocks, probably from glacial drift, that were later added to alluvial clays in relatively coarse sizes.

Evaluating the "Possibly Nonlocal Vessels" of the Ohio Valley Region

From the initial screening of thin sections from the Ohio Valley sites, 85 vessels were identified as Possibly Nonlocal (70 from Ohio plus 13 from Mann and 2 from Rutherford), and their consideration was deferred until compositional evidence from possible Southeastern sources had been compiled. The identification of Possibly Nonlocal vessels was based on two main criteria: (1) the presence of tempers other than crushed igneous or metamorphic rock grit (i.e., sand, grog, and limestone); and (2) the presence of nonlocal decoration and/or surface finishes, especially check and simple stamping. Most vessels identified as Possibly Nonlocal were characterized by both criteria; however, several vessels that were either plain or cordmarked that possessed potentially exotic tempers were also included.

During the point counting stage of the Ohio thin sections, five of the 85 vessels originally assigned to the Possibly Nonlocal category came under special scrutiny and were reevaluated. Three of these are the grog-tempered vessels from Liberty (nos. 33-149 and 33-237) and Hopewell (no. 33-154) that were identified as probable imports from the Mann site, while a grit-tempered vessel from the Mann site (no. 42) was determined to be a probable import from an undetermined Ohio site (see Table 3.3). The fifth Possibly Nonlocal vessel, from the Marietta site (no. 33-294), was reclassified as a local product because (a) its unique sandstone temper could be associated with the local bedrock and (b) its paste properties were found to be identical to those of the local pottery (see Table 2.16).

Two vessels from Seip originally included as Possibly Nonlocal were evaluated in the context of the Connestee Series at Icehouse Bottom and determined to be probable imports from that Tennessee site (see Table 5.7). With these seven vessels thus revaluated, 78 of the original 85 Possibly Nonlocal vessels remain to be considered.

Twenty-four limestone-tempered vessels (18 from Ohio, 5 from Mann, and 1 from Rutherford) plus the unique chert-tempered vessel (no. 19) from Mann were included among the original 85 Possibly Nonlocal vessels. When their paste values were compared with those of local Ohio Valley vessels, all but one (no. 33-275 from Turner) appeared to be made from local materials (see Tables 2.20 and 3.7). Further consideration of these vessels in Chapter 8 (see Table 8.4), now including body values in the discussion, confirmed the earlier finding that all but one (vessel 33-275) were locally made. Thus the status of all 24 of the limestone-tempered vessels from the Ohio Valley region plus vessel no. 19 from Mann have been addressed, reducing the number of Possibly Nonlocal vessels remaining to be considered to 53.

Not mentioned previously are ten vessels from five sites in northern Kentucky that will be considered in this chapter. These were submitted to the author for analysis because of their suspected status as Southeastern imports. Because these vessels are more or less isolated occurrences, their compositions cannot be evaluated in the context of local assemblages. On stylistic grounds, they do, indeed, appear to be intrusive: six are simple-stamped, one is check stamped, two are cordmarked, and one is plain; all appear to have sand or fine grit temper.

During the review of the 53 remaining Possibly Nonlocal vessels from the Ohio Valley plus the ten from Kentucky, it was immediately evident that an enormous amount of compositional diversity was present and that the sources of some vessels were unlikely to be determined. Instead of discussing them one vessel at a time, which would have been unnecessarily tedious, they have been subdivided into four compositional groups, three that are defined around the predominance (i.e. >50 percent) of one of the following three mineral classes: (1) Monocrystalline quartz; (2) Rock or Lithic fragments (Polycrystalline grains); and (3) Feldspars/Mafics/Micas (FMM). The fourth class, referred to as Composite, is characterized by a relatively uniform distribution of the three mineralogical classes such that no one is composed of more than 50 percent. None of these classes is, alone, sufficient to identify a specific geologic source. But, their compositions are such that their identification can provide valuable hints or limitations on where potential geological sources may be most productively sought. Each of the classes will now be considered in turn.

Monocrystalline Quartz Predominance

Thirteen of the 53 Possibly Nonlocal vessels (none of the Kentucky vessels are represented in this class) are characterized by values of monocrystalline quartz that exceed 50 percent. The geologic province that is best characterized by the dominance of monocrystalline quartz is the Coastal Plain of Georgia and Florida.

As discussed in Chapter 7, the mineralogical composition of most vessels from this region are characterized by >75 percent monocrystalline quartz inclusions along with a minimal representation of igneous and metamorphic rock grains including the minerals derived from them, such as feldspars, amphiboles and micas. Of the 26 vessels included in this study from the three Coastal Plain sites of Mandeville, Kolomoki, and Crystal River, only two (from Mandeville with 64 percent) have less than 75 percent monocrystalline quartz (see Tables 7.1, 7.2, and 7.3). Employing 75 percent monocrystalline quartz as a minimal criterion for acceptance of a vessel as being of Coastal Plain derivation, there are only three vessels in our sample that are plausible Coastal Plain derivatives, two from Seip and one from Turner (Table 10.3). A fourth vessel (33-115) from Seip could possibly be derived from Mandeville (Table 10.3). One aspect of these data stands out: assuming that the current sample of thin sectioned vessels is not seriously biased, ceramics from the Coastal Plain, especially peninsular Florida, are rare or absent from the Hopewellian sites of the Ohio Valley region.

This seems an appropriate context in which to discuss a unique vessel recovered from Leake that is a likely import from the Coastal Plain. The compositional values for this complicated stamped vessel, no. 9-270 (see Figure 6.3), are recorded in Table 10.3. By comparing these values with the two-sigma ranges recorded for Mandeville in Table 10.3 and for the monocrystalline-

Table 10.3. Bulk and Mineralogical Composition of Four "Possibly Nonlocal" Vessels from Ohio and One from Leake Characterized by Monocrystalline Quartz Predominance Compared with the Two-Sigma Ranges for the Mandeville Site

Value in bold falls beyond the two-sigma/95% confidence limits shown in the shaded cell.

		Bulk Composition				*Mineralogy*		
Site/Type	**Thin Section No.**	**% Matrix**	**% Silt**	**% Sand**	**Sand Size Index**	**% Monoc Quartz**	**% Polyc**	**% FMM**
Seip	**[n=3]**							
Plain	33-114	58	3	39	2.07	78	21	1
Punctated	33-115	63	4	33	2.13	62	**25**	13
Punctated	33-199	59	5	36	2.13	78	22	0
Turner	**[n=1]**							
Simple Stamped	33-287	57	4	39	1.45	78	2	20
Leake	**[n=1]**							
Complicated Stamped	9-270	69	3	28	1.63	93	0	7
Mandeville								
Two-Sigma Ranges	**[n=10]**	49.2-82.0	0-13.4	11.6-44.4	1.00-2.29	58.6-100	0-24.2	0-26.2

quartz-dominant petrographic class at Leake recorded in Table 10.4, it can be seen that this vessel has a Coastal Plain compositional signature. It was the only such vessel at Leake with a value for monocrystalline quartz that exceeds 76 percent.

A second possible source for vessels with a predominance of monocrystalline quartz inclusions is the Leake site, where 37 of 126 thin-sectioned vessels were assigned to a "Monocrystalline Quartz Dominant" petrographic class (Stoltman 2010; See also Table 6.4). The monocrystalline quartz values for these 37 vessels range from 51 percent to 93 percent with a mean of 62.7 percent ± 9.2. Unlike vessels from the Coastal Plain, quartz-rich vessels from Leake usually contain a generous admixture, commonly in the 20 percent to 30 percent range, of minerals derived from igneous and metamorphic rocks, especially feldspars and amphiboles, which typically dominate the FMM class.

Table 10.4 presents the bulk and mineralogical values for nine Ohio Valley vessels that compare favorably with the Leake Monocrystalline–Quartz-Dominant petrographic class. In reevaluating the vessels included in this table, it was determined that two of them—no. 33-80 from Mound City and no. 33-239 from Liberty—were best viewed as locally-made, sand-tempered vessels, not as intrusives. These vessels were originally assigned to the Possibly Nonlocal category because of their apparent sand temper. This, indeed, proved to be the case, but the relatively low incidence of sands (sand densities of <20 per 100 mm for both, which is atypical for most Southeastern vessels) and the presence of shale in both thin sections lead to the belief that both were local products.

Table 10.4 shows that the compositional values for four of the vessels (nos. 11-242, 33-308, 33-310, and no. 71 from Mann) fall fully within the 95 percent confidence limits of the Leake Monocrystalline–Quartz-Dominant petrographic class, while individual values for three vessels (33-74, 33-255, and 33-309) fall beyond those limits. The unusually large sand-size value for vessel 33-309 from Murphy III is due to a relative abundance of coarse quartz grains, including several polycrystalline ones, but since its overall compositional profile is so similar to the other two vessels from the site, it seems most parsimonious to accept it as sharing a common origin with the others. Whether the high FMM values for vessels 33-74 and 33-255 (which are due to a relative abundance of both feldspars and amphiboles) are significant (perhaps identifying unique sources?) or attributable to sampling error is indeterminable at the present time. Vessel no. 71 from Mann is deserving of special comment in that it is the only complicated stamped vessel from the site among the 11 that were thin sectioned that is not grog tempered and is regarded as a probable import from the Leake site (see Table 10.4).

Table 10.4. Bulk and Mineralogical Composition of Nine "Possibly Nonlocal" Vessels from the Ohio Valley Region Compared with the Monocrystalline-Quartz-Dominant Petrographic Class at Leake

Values in bold exceed the two-sigma/95% confidence limits shown in the shaded cells in the same column below those values.

		Bulk Composition				*Mineralogy*		
Type	**Thin Section #**	**% Matrix**	**% Silt**	**% Sand**	**Sd Size Index**	**% Monoc Quartz**	**% Polyc**	**% FMM**
Liberty	**[n=1]**							
Plain	33-239↗	**87**	6	**7**	2.00	64	27	9
Mound City	**[n=2]**							
Simple Stamped	33-74□	68	5	27	1.68	57	5	**38**
Hopewell Zoned	33-80↗	74	8	18	2.00	58	37	5
Turner	**[n=1]**							
Plain	33-255□	57	4	39	1.38	54	8	**38**
Rutherford	**[n=1]**							
Simple Stamped	11-242	73	2	25	1.49	62	20	18
Murphy I and III	**[n=3]**							
III Simple Stamped	33-308	68	5	27	1.97	53	18	29
III Simple Stamped	33-309	70	1	29	**2.51**	56	18	26
I Simple Stamped	33-310	64	11	25	1.57	54	28	18
Mann	**[n=1]**							
Complicated Stamped	#71	78	2	20	1.50	56	29	15
Leake Monoc-Quartz-Dom Two-Sigma Ranges	**[n=37]**	54.8–80.8	0–14.2	10.8–40.8	1.10–2.14	44.3–81.1	2.6–40.2	0–33.9

Note: Vessels 11-242, 33-308, 33-309, 33-310, and 71 are accepted as possible imports from Leake.

↗=*Sand-tempered vessels reclassified as local.*

□=*Vessels of undetermined origin.*

In concluding this section on vessels from the Ohio Valley region that are characterized by a predominance of monocrystalline quartz inclusions, it is reasonable to infer that the Coastal Plain of Georgia and Florida was not a major contributor. It is possible that three or four vessels were imported from the Coastal Plain, but certainly none of these vessels (cf. Tables 7.2, 7.3, and 10.1) bears the imprint of "classic" high-quartz, Coastal Plain vessels. By contrast, the Leake site of north Georgia has emerged as a potentially important source for such vessels. The compositional data presented in Table 10.4 are offered as suggestive evidence that some of the exotic vessels turning up in the Ohio Valley originated in north Georgia.

Vessels with Rock/Lithic Prominence

Polycrystalline grains may derive from sedimentary as well as from igneous and metamorphic rocks. The Appalachian Mountains are the logical source of fine-grit or sand-tempered vessels that are characterized by fragments of such igneous and metamorphic rocks as granite, gneiss, and metaquartzite, all of which occur in thin sections of Southeastern pottery. Of the 40 Possibly Nonlocal vessels plus the ten Kentucky vessels yet to be considered, nine have a predominance of polycrystalline inclusions (Tables 10.5 and 10.6). Four of these vessels were recovered from Ohio sites, whereas five are suspected of being Southeastern imports to two northern Kentucky sites. The two sites, Bentley (Henderson and Pollack 1985) and Hansen (Ahler 1992) are multicomponent (mainly Late Woodland) in nature that yielded a few Cartersville-like vessels that were provided for thin sectioning. Both sites are located in or near the floodplain of the Ohio River in Greenup County, Kentucky, directly across from the mouth of the Scioto River.

Table 10.5 records the compositional values for the four Possibly Nonlocal Ohio vessels along with two of the vessels from the Bentley site in comparison with the two-sigma ranges for the Leake metagranite petrographic class recorded at the bottom of the table. One of the latter vessels (thin section 15-1) is illustrated as Figure 3I in the Bentley site report (Henderson and Pollack 1985:147).

Two of the Ohio vessels in Table 10.5 (nos. 33-142 and 33-264) are characterized by varied lithologies of sub-rounded to rounded grains that are derived predominantly from locally available sedimentary rocks. In the case of vessel 33-142 from the Liberty Earthworks, most of the lithic fragments are derived from a sandstone with an iron-rich (i.e., opaque) cement, while Vessel 33-264 from the Turner site has several coarse grains of limestone and even some shale, all probably of local derivation. Vessel 33-244, a Zoned Incised vessel from Liberty, also appears to be a local product. Its compositional

Table 10.5. Bulk and Mineralogical Composition of Four "Possibly Nonlocal" Vessels with Prominence of Polycrystalline Grains from the Ohio Valley Region Plus Two from Bentley, Kentucky, Contrasted with the Two-Sigma Ranges for the Metagranite Petrographic Class from Leake

Values in bold exceed the two-sigma/95% confidence limits shown in the shaded cells in the same column below those values.

		Bulk Composition				*Mineralogy*		
Site/Type	**Thin Section No.**	**% Matrix**	**% Silt**	**% Sand**	**Sand Size Index**	**% Monoc Quartz**	**% Polyc**	**% FMM**
Liberty	[n=2]							
Zoned Incised	33-244☐	**87**	2	**11**	**2.88**	26	58	16
Simple Stamped	33-142☐	75	4	21	**2.60**	**37**	60	**3**
Turner	[n=2]							
Simple Stamped	33-286	64	5	31	2.38	7	**78**	15
Cordmarked	33-264☐	69	9	22	**2.59**	**39**	56	5
Bentley, KY								
Simple Stamped	15-1	67	4	29	1.56	3	65	23
Plain	15-2	67	3	30	**1.44**	6	53	41
Leake Metagranite Two-Sigma Range	[n=23]	46.5–79.7	1–12.2	14.2–47.4	1.46–2.42	2.7–27.9	24.2–77.4	3.7–64.1

Note: Vessels 33-286, 15-1, and 15-2 are accepted as probable imports from the Leake site.

☐ = *Vessels reevaluated as local Ohio products.*

Table 10.6. Bulk and Mineralogical Compositions of Three Vessels from Northern Kentucky Characterized by Metaquartzite Predominance Contrasted with the Two-Sigma Ranges for the Metaquartzite-Predominant Classes at Tunacunnhee and Leake

Site/Type	Thin Section No.	Bulk Composition % Matrix	% Silt	% Sand	Sand Size Index	Mineralogy % Monoc Quartz	% Polyc	% FMM
Bentley, KY	n=2							
Cordmarked	15-5	71	4	25	2.64⊙	0⊙	90⊙	5
Cordmarked	15-6	77◆	5	18◆	2.62⊙	18◆	82⊙	0◆
Hansen, KY	n=1							
Check Stamped	15-14	75	1⊙	24	2.54⊙	10◆	63◆	27
Tunacunnhee Metaqtzite Two-sigma Ranges◆	n=4	59.6–76.8	0.8–6.8	21.4–34.6	2.40–3.20	0–5.2	64.8–94.8	4.4–32.0
Leake Metaquartzite Two-Sigma Ranges⊙	n=4	47.4–80-6	1.4–8.2	17.0–45.4	1.75–2.27	1.7–55.3	43.5–69.5	0–35.4

Note: Vessel 15-5 is accepted as a possible import from the Tunacunnhee site.
Vessels 15-6 and 15-14 are of undetermined origin.
◆=*Values that fall outside the two-sigma limits for Tunacunnhee.*
⊙=*Values that fall outside the two-sigma limits for Leake.*

profile does not fit at all with the Southeastern pattern. Indeed, its sand inclusions are so sparse (bulk sand percent of 11 percent, with a sand density of only 8.8 grains / 100 mm) and generally coarse grained (bulk size=2.88) that an idiosyncratic Ohio origin seems likely. These three vessels, then, are now reconsidered to be local manufactures.

In contrast to these reevaluated vessels, simple-stamped vessel 33-286 from Turner is unquestionably an import from the Appalachian region. The majority of its sand-size inclusions derive from a rock of granitic composition, with polycrystalline quartz, feldspars, and micas all present both individually and in polycrystalline grains. Yet unlike Ohio, where granite grains are generally coarser, the rock fragments in this vessel are relatively fine grained (note the bulk sand-size index of 2.38 in Table 10.5). The polycrystalline grains show several signs of metamorphic alteration, including foliation or layering, similar to metagranite grains at Leake. That this vessel derives from the Leake site is considered highly likely, as can also be seen from how closely the compositional values agree with vessels of the Leake Metagranite petrographic class (see Table 10.5).

The two vessels from Bentley also have compositional profiles similar to that of the Metagranite petrographic class at Leake, with but one minor departure for vessel 15-2 (see Table 10.5). The closest counterpart to these vessels within our current sampling universe clearly occurs at the Leake site.

Finally, three vessels from northern Kentucky, two from Bentley and one from Hansen, are discussed here because of apparent similarities to the Quartzite-rich vessels from Tunacunnhee and Leake. Vessel 15-14 from the Hansen site is check stamped and was thought to be of special interest for this study. The compositional data for these three vessels are recorded in Table 10.6 along with the relevant two-sigma ranges for Tunacunnhee and Leake vessels. From Table 10.6, it can be seen that all compositional values for cordmarked vessel 15-5 fall fully within the two-sigma ranges for the Metaquartzite-prominent class at Tunacunnhee. If these data can be accepted at face value, this is the only vessel in our current sample that could be an export from this north Georgia site. The other two quartzite-rich vessels in Table 10.6 do not appear closely related to either of the Quartzite-rich classes at Leake or Tunacunnhee. No source can be suggested for these vessels at present.

Vessels with Feldspar/Mafic/Mica (FMM) Prominence

This is the largest of the four mineral classes discussed in this chapter and includes 23 of the 36 Possibly Nonlocal vessels yet to be considered plus the remaining five Kentucky vessels. Two of the 23 were discussed previously in Chapter 5 where they were identified as Connestee Simple-Stamped vessels intrusive to Seip from Icehouse Bottom (see Table 5.7 and Figure 5.8), so they are not discussed further here. The remaining vessels in this mineral class will be considered in two groups, the first consisting of six Possibly Nonlocal vessels that are recorded in Table 10.7.

These six vessels owe their high FMM values to an abundance of amphiboles, typically with an admixture of feldspars. As a result, the two-sigma ranges for two sites at which amphiboles are especially abundant, Garden Creek and Leake, are recorded for comparative purposes at the bottom of Table 10.7. In this table individual values that fall beyond the two-sigma ranges of one or both sites are identified with symbols that identify which of the sites is involved as follows: ◆ for Garden Creek and ⊙ for Leake. As can be seen from Table 10.7, all compositional values for two vessels—No. 33-186 from Seip and No. 80 from Mann—fall completely within the two-sigma ranges for the Amphibole-rich petrographic class at Leake.

The other four vessels in Table 10.7 (Nos. 33-72, 33-75, 78, and 82), with the exception of slightly elevated values for monocrystalline quartz, also have compositional values that conform closely to those of the Leake Amphibole-rich petrographic class. By contrast, their overall compositional profiles do not conform well to those of Garden Creek. It seems reasonable to see these compositional similarities to Leake as suggestive evidence that all six of these vessels were derived from that site. There is certainly no other source that can be reasonably suggested for these vessels within the currently available thin section sample.

The second group of FMM-rich vessels, those characterized by "Grit A," is the largest group of imported vessels recovered from the Ohio Hopewell sites. No fewer than 17 vessels of this group are documented from five Ohio sites, and, as can be seen from Table 10.8, they are surprisingly uniform in composition. Only two of the vessels have compositional values that fall outside the two-sigma ranges for the entire group of 17, and these values are so minor that sampling error is almost certainly the reason. Fifteen of the 17 vessels are simple stamped, while two are plain (see Table 10.8).

Identifying the presence of Grit A, so-called because of the suspicion that the visible inclusions were derived from finely crushed igneous rocks, is relatively straightforward due to the prominence of microcline, a potassium-rich feldspar that is characterized by distinctive tartan twinning when viewed under crossed polars (Figure 10.1). Amphiboles are co-dominant with microcline in thin sections with Grit A, so that the FMM values are usually greater than 50 percent (see Table 10.8). These amphiboles (e.g., mostly cummingtonite, gedrite, and tremolite) differ in appearance from hornblende under plane polars, being colorless (or nearly so) and weakly pleochroic. Pla-

Table 10.7. Bulk and Mineralogical Compositions of Six "Possibly Nonlocal" Vessels from the Ohio Valley Region Characterized by Feldspar/Mafic/Mica (FMM) Predominance Contrasted with the Two-Sigma Ranges for the Connestee Series at Garden Creek and the Amphibole-Rich Class at Leake

		Bulk Composition				*Mineralogy*		
Site/Type	**Thin Section No.**	**% Matrix**	**% Silt**	**% Sand**	**Sand Size Index**	**% Monoc Quartz**	**% Polyc**	**% FMM**
Seip								
Simple Stamped	33-186	65	6	29	1.62	14	18◆	68◆
Mound City								
Plain	33-72	74◆	5	21◆	1.58	39	9◆	52
Simple Stamped	33-75	65	7	28	1.50	41⊙	6◆	53
Mann								
Fine Simple Stamped	No.78	72◆	4	24	1.82	35◆⊙	6◆	59
Fine Simple Stamped	No.80	67	1◆	32	1.72	30◆	15◆	55
Fine Simple Stamped	No.82	62	5	33	1.90	35◆⊙	13◆	52
Garden Creek◆ Connestee Series	n=12	47.4-71.6	2.4-10.0	24.0-44.8	1.61-2.25	6.6-19.4	22.2-75.8	9.6-66.4
Leake⊙, Amph-Rich Two-Sigma Range	n=14	57.1–78.3	0–13.2	12.5–40.1	1.22–1.90	0–34.4	2.5–41.7	34.7–86.7

Note: All vessels above are accepted as possible imports from the Leake Site.
◆=*Values that fall outside the two-sigma limits for Garden Creek.*
⊙=*Values that fall outside the two-sigma limits for Leake.*

Table 10.8. Bulk Composition and Mineralogy for Ohio Vessels with "Grit A"

Values in bold fall outside of the relevant two-sigma ranges shown in shaded cells.

Site (Type)	Thin Section No.	*Bulk Composition* % Matrix	% Silt	% Sand	Sand Size Index	*Mineralogy* % Monoc Quartz	% Polyc	% FMM
Seip (Simple Stamped)	33-110	61	4	35	1.58	1	6	93
Seip (Simple Stamped)	33-111	57	9	34	1.64	1	15	84
Seip (Simple Stamped)	33-190	63	3	34	**1.35**	0	16	84
Seip (Simple Stamped)	33-192	**49**	9	**42**	1.68	0	42	58
Seip (Simple Stamped)	33-195	57	4	39	1.67	0	28	72
Mean ±1 Std Deviation	**[n=5]**	57.4±5.4	5.8±3.0	36.8±3.6	1.58±.14	0.4±0.6	21.4±13.9	78.2±13.5
Turner (Simple Stamped)	33-262	73	1	26	1.81	27	32	41
Turner (Simple Stamped)	33-263	68	10	22	1.62	3	25	72
Turner (Simple Stamped)	33-267	68	3	29	1.60	21	42	37
Turner (Simple Stamped)	33-268	65	8	27	1.60	15	22	63
Turner (Simple Stamped	33-281	62	6	32	1.62	28	22	50
Turner (Simple Stamped)	33-285	62	6	32	1.63	30	34	36
Turner (Plain)	33-277	64	10	26	1.51	16	20	64
Turner (Plain)	33-282	73	3	24	1.56	25	23	52
Mean ±1 Std Deviation	**[n=8]**	66.9±4.4	6.0±3.2	27.1±3.5	1.62±.09	20.9±9.0	24.6±5.5	54.5±12.1
Mound City (Simple Stamped)	33-73	66	4	30	1.74	0	18	82

Mound City (Simple Stamped)	33-76	71	1	28	1.68	0	19	81
Mean ±1 Std Deviation	**[n=2]**	68.5±3.5	2.5±2.1	29.0±1.4	1.71±.04	0±0	18.5±0.7	81.5±0.7
Newark (Simple Stamped)	33-232	60	3	37	1.72	4	30	66
Grimes (Simple Stamped)	33-107	61	7	32	1.69	1	42	56
Mean ±1 Std Deviation	**[n=17]**	63.5±6.2	5.4±3.0	31.1±5.4	1.63±.10	10.3±12.0	24.3±9.5	65.4±16.1
Ohio Two-Sigma Ranges	**[n=17]**	51.1-75.9	0-11.4	20.3-41.9	1.43-1.83	0-34.3	5.3-44.3	33.3-97.6

Note: All vessels of undetermined origin.

10.1. Photomicrograph of Simple-Stamped Vessel 33-110 with "Grit A" taken at 10X magnification under crossed polars. Note abundance of microcline (with tartan twinning) and amphiboles [A] (with relatively high relief). Scale: large microcline grain lower left [Mi]= .675 mm.

gioclase and micas are rare to nonexistent, while the occurrence of quartz is highly variable, ranging from trace amounts to percentages as high as 30 percent (see Table 10.8). It is possible that two sources of Grit A exist, one with 15–30 percent quartz, which occurs only at Turner, and one with <5 percent quartz, which occurs at the other four Ohio sites (see Table 10.8). The vast majority of inclusions are monocrystalline, but rock particles such as polycrystalline quartz and amphibolite do occur sparsely. One final distinctive feature of these vessels is the fine-grained nature of the mineral inclusions—the mean sand size index is $1.63 \pm .10$—and their great abundance, i.e., mean sand volume$=31.1 \pm 5.4$ (see Table 10.8) and a mean sand density value of 29.8 ± 4.6 grains per 100 mm.

The mineral suite that characterizes these vessels is clearly derived from the Appalachian Mountains (perhaps North Carolina), but so far no credible source has emerged. Interestingly, there are several occurrences of Grit A in the Southeast, but in all known cases these are almost certainly intrusive vessels just as in Ohio (Table 10.9).

One vessel with Grit A was recovered at Garden Creek (simple-stamped vessel 31-15). By comparing the compositional values for this vessel with those

Table 10.9. Bulk Composition and Mineralogy for Vessels from Southeastern Sites with "Grit A"

Values in bold fall outside of the relevant two-sigma range shown in the shaded cell.

		Bulk Composition				*Mineralogy*		
Site (Type)	**Thin Section No.**	**% Matrix**	**% Silt**	**% Sand**	**Sand Size Index**	**% Monoc Quartz**	**% Polyc**	**% FMM**
Garden Creek, NC	[n=1]							
Simple Stamped	31-15	61	8	31	1.48	22	20	58
Icehouse Bottom, TN	[n=2]							
Plain	40-14	63	7	30	**1.32**	22	12	66
Plain	40-16	64	3	33	1.76	3	28	69
Camargo, KY								
Simple Stamped	15-18	62	4	34	**1.38**	1	23	76
Panther Rock, KY	[n=1]							
Simple Stamped	15-20	59	5	36	1.62	11	11	78
Bentley, KY	[n=2]							
Simple Stamped	15-3	64	7	29	**1.32**	13	11	76
Simple Stamped	15-4	67	2	31	1.80	8	37	55
Ohio 2-Sigma Ranges	**[n=17]**	51.1–75.9	0–11.4	20.3–41.9	1.43–1.83	0–34.3	5.3–44.3	33.3–97.6

All vessels of undetermined origin.

of the Connestee series at Garden Creek (see Table 5.2) and the group of 17 Grit A vessels from Ohio (Table 10.9), its affinities are unambiguously with the latter rather than the former. Two vessels with Grit A are among the thin sections from Icehouse Bottom (plain vessels 40-14 and 40-16). Once again compositional comparisons of these vessels with the local Connestee series at Icehouse Bottom (see Table 5.5) and the 17 Grit A vessels from Ohio (see Table 10.9), reveals clearly that they are much more closely related to the latter than to the former.

Four simple-stamped vessels from three sites in northern Kentucky—Camargo, Panther Rock, and Bentley—were also characterized by Grit A (Table 10.9). The vessels from Camargo (15-18; Fenton and Jeffries 1991:52, Figure 19) and Panther Rock (15-20) are nearly complete simple-stamped jars with tetrapodal bases. The compositional values for these four vessels, with the exception of two cases of sand-size indices being a bit smaller than the two-sigma range for the 17 Ohio vessels, all fall within the two-sigma ranges for the 17 Ohio vessels (see Table 10.9). Since fine-grained inclusions are the rule for this group of vessels, that these three (counting also vessel 40-14 from Icehouse Bottom) were slightly finer grained should come as no surprise.

Vessels of Composite Composition

The thirteen remaining unevaluated Possibly Nonlocal vessels from Ohio Valley sites have compositional profiles that approximate those of the Composite petrographic class at the Leake site. Table 10.10 records the compositions of these 13 vessels for comparison with the two-sigma ranges for the Composite petrographic class at Leake recorded at the bottom of the table. Also included in Table 10.10 are two additional simple-stamped vessels, one with a tetrapodal base that was recovered at the Amburgey site in Montgomery County, Kentucky (Richmond and Kerr 2005:82–83) and the other from Crystal River (cf. Table 8.3). The Amburgey vessel is stylistically similar to those from Camargo and Panther Rock. This Cartersville-looking jar was recovered from a Middle Woodland feature that also contained two copper earspools. Cordage from one of the earspools produced a two-sigma calibrated C-14 date of A.D. 40 (110) 230 (Richmond and Kerr 2005:81).

Because the Composite class lacks the specificity of the other mineralogical classes, one is left to wonder whether it has enough integrity to be considered more than a residual grouping of vessels that could be represented at virtually any site. It is noteworthy that the 15 vessels in Table 10.10 have a surprising amount of compositional uniformity. The only values that fall beyond the two-sigma ranges for the group involve five vessels with sand size indices that exceed 2.09, plus one of those five (33-152) with an unusually low incidence of monocrystalline quartz (see Table 10.10). In evaluating this group of vessels, four points deserve emphasis: (1) all bulk composition values other than the sand size index fall within the relevant two-sigma ranges for the Leake site; (2) only one value within the three mineralogical classes (11 percent Monocrystalline Quartz for vessel 33-152) falls outside the two-sigma ranges for those classes; (3) underlying the apparent general mineralogical uniformity are a suite of specific minerals that make up these classes—feldspars, amphiboles, polycrystalline quartz, and micas—all of which derive from rocks of metagranite or metaquartzite composition; and (4) the composite class predominates (47 of 126 vessels) at the Leake site (see Table 6.4). It may be premature to accept vessels with such varied mineralogical composition as indicative of a single source, but there can be no reasonable doubt that these vessels ultimately reflect the geology of the Appalachian Mountains. Whether or not their presence in the Ohio Valley region can be attributed to Leake alone is certainly debatable, but until an alternative credible source is identified, a Leake-derivation is currently the most reasonable hypothesis available for their occurrence in these various sites.

Concluding Observations

This chapter has focused on identifying, insofar as possible, credible sources for 95 vessels at certain sites (70 from Ohio Hopewell sites, 13 from Mann, 2 from Rutherford, plus 10 from northern Kentucky) suspected of having been transported from the Southeast to the Ohio Valley region. During the early stages of the reevaluation process, six vessels (3 from Liberty, and one each from Mound City, Turner, and Marietta) were reclassified as local as were 23 of the 24 limestone-tempered vessels and the unique chert-tempered vessel from Mann, leaving a residue of 65 Possibly Nonlocal vessels for further consideration as probable imports. In the final analysis of these 65 vessels, sources for 25 could not be determined, resulting in a total of 40 vessels for which an external source could be suggested. Table 10.11 summarizes the inferred sources for these 40 vessels plus 15 other vessels that were regarded as imports to 5 Southeastern sites.

Without question, the most surprising and interesting outcome of the evaluation of pottery vessels that circulated among sites of the Ohio Valley region and the Southeast is the prominence of two sites, Leake and Mann. Pottery vessels have never been considered among the items that were distributed within the Hopewell Interaction Sphere (e.g., Seeman 1979), so it was unanticipated that pottery vessels apparently made at Leake and Mann were identified from several sites that were known participants in the Hope-

Table 10.10. Bulk and Mineralogical Composition of 13 "Possibly Nonlocal" Vessels from Ohio Valley Sites Plus One Vessel Each from Amburgey and Crystal River Compared with the Composite Petrographic Class at Leake

Values in bold exceed the two-sigma ranges shown in the shaded cells below those values.

		Bulk Composition				*Mineralogy*		
Site/Type	**Thin Section No.**	**% Matrix**	**% Silt**	**% Sand**	**Sd Size Index**	**% Monoc Quartz**	**% Polyc**	**% FMM**
Hopewell	**[n=3]**							
Plain	33-152	69	7	24	**2.54**	**11**	42	47
Plain	33-153	70	5	25	2.08	46	36	18
Simple Stamped	33-158	72	6	22	1.88	33	36	31
Mound City	**[n=1]**							
Simple Stamped	33-77	70	4	26	1.62	25	29	46
Seip	**[n=3]**							
Simple Stamped	33-188	67	4	29	1.58	38	28	34
Simple Stamped	33-187	56	10	34	1.69	43	27	30
Simple Stamped	33-189	67	10	23	1.51	49	17	34
Turner	**[n=3]**							
Plain	33-254	65	8	27	**2.35**	46	48	6
Simple Stamped	33-283	60	8	32	1.60	41	20	39
Simple Stamped	33-284	67	16	17	1.56	33	27	40
Newark	**[n=1]**							
Simple Stamped	33-233	69	1	30	**2.38**	21	47	32

Mann	**[n=2]**							
Fine Simple Stamped	79	76	6	22	1.60	25	39	36
Fine Simple Stamped	81	65	3	32	2.00	42	14	44
Amburgey, KY	**[n=1]**							
Simple Stamped	15-19	69	3	28	**2.36**	38	34	28
Crystal River, FL	**[n=1]**							
Simple Stamped	8-12	63	5	32	**2.57**	31	39	30
Leake Composite, Two-Sigma Ranges	**[n=47]**	55.4-80.2	0-19.1	11.0-35.0	1.13-2.09	19.4-55.8	13.1-50.7	9.7-51.3

Table 10.11. Numbers of Vessels Regarded as Foreign Recovered at Specific Sites Cross-Tabulated with the Suspected Sites from Which Those Vessels Originated

		Sites Exporting Vessels							
Sites Importing Vessels	**N**	**Icehouse Bottom**	**Garden Creek**	**Coastal Plain**	**Tunacunnhee**	**Leake**	**Mann**	**Ohio**	**Unknown**
Ohio Valley Sites									
Mann	7					6		1	—
Liberty	2						2		—
Mound City	6					3			3
Hopewell	4					3	1		—
Seip	14	2		3		4			5
Turner	15	1		1		4			9
Newark	2					1			1
Murphy	3					3			—
Grimes	1								1
Rutherford	1					1			—
Amburgey, KY	1					1			—
Bentley, KY	6				1	2			3
Panther Rock, KY	1								1
Camargo, KY	1								1
Hansen, KY	1								1
Subtotals	65	3		4	1	28	3	1	25

Southeastern Sites									
Icehouse Bottom	6		2					2	2
Leake	4	1		1			2		—
Garden Creek	2						1		1
Crystal River	1					1			—
Pinson Mounds	8	1				3	1		3
Subtotals	21	2	2	1		4	4	2	6
Grand Totals	86	5	2	5	1	32	7	3	31

well Interaction Sphere. Vessels from Mann, characterized by their distinctive grog temper, were recovered not only from two Ohio sites, Liberty and Hopewell, but also from such widely separated sites as Garden Creek in North Carolina, Pinson Mounds in west Tennessee, and Leake in north Georgia.

That Mann potters were interacting with Swift Creek artisans has long been believed, but that most of its uniquely abundant, Swift Creek–like complicated stamped pottery vessels were made locally, not imported (Ruby and Shriner 2005) confounded understanding of that connection. The simplest mechanism, direct exchange of pottery vessels, was apparently rare. The recent excavations at the Leake site have shed important new light on the relationship of Mann to the Swift Creek world by providing firm evidence of a direct connection between the two sites. As reported in Chapter 3, two vessels recovered at Leake, one rocker stamped the other complicated stamped, bear the unmistakable compositional signature of the Mann site. Moreover, the five fine-simple-stamped vessels from Mann that Ruby and Shriner (2005) properly identified as imports from Appalachia, plus one complicated stamped vessel, have now been accorded a more specific source attribution (i.e., Leake; see Tables 10.7 and 10.10). With Leake revealed as an active partner with Mann, the two sites can be productively viewed as poles along an important axis of cultural interaction between the Appalachian and Ohio Valley regions (Keith 2010b).

11
Summary and Conclusions

A major goal of this study is to refine and apply several quantitative indices to the petrographic analysis of ceramic bodies. Traditionally primarily a qualitative technique, it is intended that these indices will enhance the potential of petrography to yield valuable insights into the behavior of past potters. For example, by identifying the natural resources present in ceramic bodies, useful inferences can be made concerning how the local environment was being exploited. More important, by reliably identifying the mix of mineral resources involved in the production of ceramic bodies, valuable insights may be gained concerning decisions and practices of the past artisans themselves.

Turning now from the methodological to the substantive contributions of this volume, I summarize the four most important findings.

I. Characterizing the Composition of Ohio Hopewell Ceramics

The main substantive contribution of this volume is the compilation of compositional data on 586 pottery thin sections from 31 Middle Woodland sites extending from the central Ohio Valley to the Gulf coast of Florida and from North Carolina to western Tennessee. The initial focus of the volume was upon the Hopewell sites of southern Ohio, with the primary objective to establish a compositional baseline for Ohio Hopewell pottery. In the process of establishing this compositional baseline, several vessels were recognized as stylistically as well as compositionally anomalous. These were set aside as "Possibly Nonlocal" until the local signatures had been established for the Ohio sites and compositional data for comparative purposes could be assembled from Southeastern sites suspected of being sources of the exotic vessels.

Scioto and Hopewell Wares, Two Recipes

There were three findings pertaining to Ohio Hopewell pottery that standout as especially noteworthy. Foremost among them was the discovery that

Scioto and Hopewell wares of the major Hopewell centers of the Chillicothe region were made with different recipes. The former (always undecorated) was endowed with coarser and more abundant crystalline-rock temper in contrast to the decorated Hopewell ware, which has generally lesser amounts and finer grades of temper. These data are compiled in Table 2.13 for the five main sites of the Chillicothe region. Student's t-tests for significance of the differences between the two wares pertaining to temper size and temper amount were both significant at the .001 level. From this it is reasonable to infer that the compositional differences were intentional and reflect the experience and wisdom of the potters to use different recipes for wares that served different functions: Scioto ware for cooking, storage, and other utilitarian purposes and Hopewell ware for serving, ritual, and ceremonial purposes. The greater sizes and amounts of temper in Scioto ware were surely to provide mechanical and thermal shock resistance to breakage that would have accrued during everyday handling, which would have included recurrent heating and cooling associated with cooking.

Grog Temper at Tremper

The second finding with respect to Ohio Hopewell pottery was the surprising discovery that the local pottery at the Tremper site was tempered with grog. This finding is unique for a Ohio Hopewell site, but is in conformity with a more general pattern that sees grog temper used at other Middle Woodland sites in the lower Ohio valley, namely the Mann site in southern Indiana and 11Mx109 in southern Illinois.

The Turner Series and Limestone Temper, Local Ohio Products

The third notable finding with respect to Ohio Hopewell pottery is that several check and simple-stamped vessels of the Southeastern series (i.e., the Turner types) were, indeed, locally made, as were most limestone-tempered vessels recovered in the Ohio sites. These findings confirm the views expressed by Prufer (1968; see also Prufer and McKenzie in Prufer 1965:18–59). Interestingly, with the exception of a check stamped vessel from the Hansen site in Kentucky (which could not be identified to a source but was probably of Southeastern derivation), all of the check stamped vessels from Ohio Valley sites were locally made.

II. Distinguishing Southeastern from Ohio Valley Pottery

The "sandy feel" or texture of stylistically anomalous sherds believed to be of Southeastern origin (especially those with simple stamping) recovered from

Ohio Hopewell sites has long been recognized (e.g., Prufer 1968:8–9). In most cases two properties permitted the reliable identification of these sherds in comparison to the local Hopewell pottery: (1) a greater density of mineral inclusions that were (2) generally finer grained. In an attempt to provide an objective basis for distinguishing these exotic sherds, the following two "compositional profiles" were defined based upon the thin section analyses. On the one hand Ohio Hopewell ceramics (Tremper excepted), which are usually made of alluvial sediments tempered with coarse igneous and metamorphic rock fragments, have the following compositional profile: bulk sand-size indices (which include temper as well as natural sand grains) exceed 3.00 and the number of sand-size grains (i.e., sand density) is normally below 20 per 100 1-mm counts. As is the case with most "rules," there are exceptions. Notably, local Hopewell ware generally has sand-size indices that fall between 2.30 and 3.00 (Table 10.1). By contrast, Southeastern vessels, with rare exceptions, have bulk sand size indices between 1.00 and 2.00 and sand density indices of greater than 20 per 100 1-mm counts, often reaching values greater than 30 (see Table 10.2). A notable exception to the "2.00-maximum sand-size rule" in the Southeast occurs in north Georgia where quartzite was present in some vessels at Tunacunnhee and Leake (see Table 10.2).

More than distinguishing Ohio Hopewell from Southeastern sherds, these compositional profiles are a reflection of two distinctly different ceramic-making traditions: one (Ohio Hopewell) involving tempering alluvial clays with coarse grains of crystalline rocks (derived from glacial till) in relatively low densities in contrast to the other (Pan-Southeastern) involving the use of finely crushed or carefully selected grains of local minerals and rocks in relatively high densities. What is most significant about the Southeastern practice is that it occurs throughout this culturally diverse and geologically complex region.

III. Accounting for the Dispersal of Pottery Vessels among Sites That Participated in the Hopewell Interaction Sphere

An unanticipated finding of this study, after the results had been tabulated, is that a surprisingly large number of pottery vessels had been recovered from Middle Woodland sites across the Southeast (including the lower Ohio Valley) that were not locally manufactured. Indeed, from among the 586 thin-sectioned vessels in this study, 86 were identified as intrusive to the site of their recovery.

In an effort to explain the significance of the widespread occurrence of "trade" vessels within Middle Woodland contexts across the Southeast, let me open with the assertion that trade or exchange in most cases does *not* account for their observed distribution. It is likely that these vessels were not interaction sphere goods *per se*, which are mainly exotic and distinctive raw materials and artifacts; rather, it is more likely that they were part of the personal gear of travelers engaged in interaction sphere activities. Support for this view derives from two observations.

First, the pottery vessels involved, with few exceptions (e.g., the red-slipped vessel from Pinson Mounds and the burnished vessel from Garden Creek), are types that were everyday, utilitarian jars (e.g., simple stamped, cordmarked, and plain) rather than special-purpose containers. If the pottery vessels themselves were media of exchange, it is expectable that they would be endowed with properties characteristically associated with ritual or ceremonial uses, such as distinctive shapes or unusual decoration. But, of course, it could have been the contents, rather than the vessels themselves, that were exchanged, which is a credible view, but one that is not currently testable in the archaeological record. Perhaps someday identification of residues from imported sherds could be employed to address this issue.

Second, the geographic distribution of sites producing the anomalous or exotic vessels is decidedly asymmetrical, which is inconsistent with direct reciprocity ordinarily associated with "trade." Of the 55 vessels for which source identifications have been proposed (this excludes the 31 imports from unidentified sources listed in Table 10.11), sites in the Ohio Valley exported but 10 (7 from Mann and only 3 from the Ohio Hopewell centers, one of which went to Mann) while 45 vessels were exported from Southeastern sites (see Table 10.11). These figures are even more unbalanced if the 31 vessels for which source identifications could not be determined are included because it is highly probable that all of these (21 of which were tempered with Grit A) were Southeastern derivatives. If exchange for vessel contents was reciprocated directly with materials from the Ohio Valley, there is currently relatively little archaeological evidence for it.

The simplest explanation for these observations is that Southeasterners were recurrent, short-term visitors to the northern centers. In making such visits they transported pottery vessels with them to prepare meals during their travels and while resident in or near the northern centers. Leaving behind vessels as gifts or due to breakage in use would reasonably account for the presence of "trade" vessels in the northern centers. The same processes would have been in operation in reverse, but the volume of traffic from the Ohio Valley region to the south seems clearly to have been on a much smaller scale. Under this view, the pottery vessels would not be seen as tribute nor as commodities associated with economic transactions. Rather, they are best conceived as accessories intended primarily to facilitate the safe journeys of the travelers.

A byproduct of such visits would be inevitable interaction between potters of the two regions. Such interaction surely accounts for the occurrence of pottery types in the Ohio centers that are stylistically exotic (i.e., Turner Simple Stamped A and Turner Check Stamped) but made of local materials. Indeed, all 24 of the Turner-series vessels recovered from Ohio Hopewell sites that were thin sectioned were made of local materials.

There are several possible reasons or motives for Southeasterners to sojourn to the Hopewell centers of the Ohio Valley region. Carr (2005:579–580) provides an impressive list of ten: (1) vision and power questing; (2) pilgrimage to a place of power in nature; (3) travels of medicine people to heal the sick, or sick people to seek healing; (4) buying, selling, or learning ceremonial rites; (5) spirit adoption; (6) interregional intermarriage; (7) pilgrimage to a ceremonial center; (8) valuables exchange among elites; (9) travels to a center of learning to gain esoteric knowledge; and (10) elite-orchestrated transference of religious cults to facilitate supralocal exchange. All of these, except perhaps No. 2, are credible reasons for Southeasterners to visit the northern centers.

The main problem in attempting to discriminate among these alternatives is identifying archaeological correlates for the underlying motives, which are, after all, nonmaterial. It is likely that the purveyors of the pottery vessels were more than simple visitors to the northern sites. Rather, whether acting individually or in behalf of others (e.g., leaders, elites, or family groups), it is likely that they were on "missions" that were motivated by one or more of the above alternatives and that the pottery vessels are best regarded as part of the gear of the travelers, not their primary wares. Unfortunately, then, the effectiveness of the exotic vessels to inform about the real objectives of the Southeastern visitors to the Hopewell centers is limited. Whatever the reasons of the travelers, the northern Hopewell centers certainly exerted a substantial attraction for Southeasterners.

IV. The "Leake-Mann axis of interaction"

Confirmation of what might be referred to as the "Leake-Mann axis of interaction," first recognized by Scot Keith (Keith 2010b), is one of the more important outcomes of this study. The former site, which is published as a detailed contract report (Keith 2010a), and the latter, which for whatever reason, remains generally under appreciated in the literature of Eastern North America, now appear to be major "players" in the broad network of interacting Hopewellian societies. The lack of prominence of the two sites in the archaeological literature is surely due in part to the absence of excavated mound burials, which is the primary recovery context for Hopewell Interaction Sphere goods in the East. Nonetheless, Leake and Mann are now documented as *the* two most active producers of pottery vessels that circulated among Middle Woodland sites of the Eastern Woodlands south of the Ohio and east of the Mississippi Rivers.

In the case of Mann, its distinctive grog-tempered pottery vessels have been identified from five Middle Woodland sites extending from Garden Creek in North Carolina to Pinson Mounds in west Tennessee and from the Liberty and Hopewell sites in Ohio to Leake in north Georgia. With respect to Leake, if the evidence reported herein can be accepted at face value, this north Georgia center has emerged as *the* most active participant in the widespread dispersal of pottery to sites within the orbit of the Hopewell Interaction Sphere. At least 32 vessels manufactured at the Leake site (25 of which are simple stamped) found their way to the following Middle Woodland sites: Hopewell (n=3); Mound City (n=3); Seip (n=4); Murphy (n=3); Newark (n=1); Turner (n=4); Mann (n=6); Rutherford (n=1); Amburgey (n=1); Bentley (n=2); Crystal River (n=1); and Pinson Mounds (n=3). That is an astounding 12 sites extending from the Florida Gulf Coast to the Ohio Valley (see Table 10.11).

Especially important within these data pertaining to the circulation of ceramic vessels is evidence of direct two-way interaction between the Leake and Mann sites. On one side, this evidence consists of two grog-tempered sherds at Leake that are compositionally indistinguishable from Mann pottery. One of these sherds is rocker stamped (clearly anomalous at Leake) and the other is complicated stamped. Talk about bringing coals to Newcastle, for the latest of two major Middle Woodland components at the Leake site "is dominated by Swift Creek Complicated Stamped and Cartersville Simple Stamped pottery" (Keith 2010a:457)!

Going in the reverse direction, the findings in this study (see Chapter 10) suggest that the five fine simple-stamped sherds identified as generally Southeastern imports to Mann (e.g., Ruby and Shriner 2005:568) were derived from the Leake site. There is certainly no realistic alternative source for these sherds, which conform to the type Cartersville Simple Stamped that is so common at Leake. In addition a complicated stamped sherd with quartz-dominant inclusions (vessel no. 71 at Mann) also appears to be derived from Leake. This vessel, interestingly, is the only recorded example of a complicated stamped sherd from Mann that is not grog tempered.

One of the persistent enigmas about the Mann site is how to explain the presence of literally hundreds of Swift Creek–like complicated stamped sherds (e.g., Kellar 1979:103), the most at any site north of the Swift Creek heartland; whereas, only 13 such sherds have been reported from all of the Ohio Hopewell sites combined (Prufer 1968:14). Compounding the dilemma is evidence that the preponderance of the complicated stamped pottery was locally made at Mann (Ruby and Shriner 2005).

There is now a plausible explanation for this enigma, especially when the data from Pinson Mounds are incorporated into the discussion. For, the two sites, Mann and Pinson Mounds, share a pattern of culture contact with the Swift Creek world, namely, that carved paddles rather than pots were transported to sites located well beyond the Swift Creek heartland. Who actually used those paddles is a more complicated issue, for it is possible that paddle-carrying visitors made the pottery (the explanation favored here) or that local people made the pottery using paddles that had been procured or copied through cultural interaction. Either alternative would account for the complicated stamped pottery made of local materials that was recovered at Mann and Pinson Mounds.

Beside evidence of cultural interaction between the occupants of the Leake and Mann sites, there are two important similarities between the sites that contrast markedly with the Ohio Hopewell centers. First and foremost, both sites were intensively occupied, with extensive middens and features indicative of a substantial resident population (Ruby 2006:198–199; Keith 2010:59–294), quite in contrast to the vacant-ceremonial-center/dispersed-hamlet model advocated by Prufer (1964:71) for the Ohio centers and subsequently forcefully supported by more recent research (e.g., Dancey and Pacheco 1997; see also Carr 2008). Second, neither site has yet produced evidence of classic Hopewell mound burials.

Both Leake and Mann appear to have been major regional centers that served multiple functions. From the ceramic evidence, it appears that a significant part of the resident population of both sites was comprised of foreigners. This inference is based upon the abundance of stylistically foreign pottery vessels made of local materials. This abundance is notably far greater than occurs at the Ohio Hopewell centers where interaction with short-term visitors would better account for the presence of vessels with foreign styles made of local materials.

In the case of Leake one cannot but be impressed by the enormous stylistic diversity of a ceramic assemblage composed almost entirely of Appalachian highlands minerals without postulating face-to-face interaction among people from diverse Southeastern regions, extending from the Coastal Plain to the Lower Mississippi Valley (refer to Figure 6.3, Table 6.3, and Keith 2010a). It is likely that the site served as a focal point, perhaps a point of embarkation to the north, for Southeasterners traveling to the Hopewell centers. Mann, in similar fashion, has a unique abundance of stylistically foreign (i.e., complicated stamped) pottery made of local materials. How long such foreigners resided at the sites is unknown, but it must have been long enough to "use up" their own vessels, which were then replaced with pottery made of local materials. Presumably at least some of these nonlocal people were temporary visitors going to and coming from the Ohio Valley.

The Leake site certainly appears to have been a place where people from across the Southeast congregated. Considering the distances involved between the Leake and Mann sites (at least 500 km overland and over 600 km via the Tennessee River to the Ohio), travelers obviously needed information about routes, risks, and destinations before setting off. Considering the evidence for direct contact between the two sites, at least some residents of the Leake site must have been familiar with the Mann site and how to get there. One of the more intriguing aspects of the Leake site is that it was the source of the preponderance of identifiable "trade" vessels (32 of 55; see Table 10.11) recovered from Hopewell-related sites. Based on these data one could reasonably infer that Leake residents served as outfitters and guides for Southeasterners seeking to make their way to the famous Hopewell centers of the north.

References Cited

Adams, William R.

1949 *Archaeological Notes on Posey County, Indiana.* Indiana Historical Bureau, Indianapolis.

Ahler, Steven R.

1992 The Hansen Site (15Gp14): A Middle/Late Woodland Site Near the Confluence of the Ohio and Scioto Rivers. In *Cultural Variability in Context, Woodland Settlements of the Mid-Ohio Valley*, edited by Mark F. Seeman, pp. 30–40. MCJA Special Paper No. 7, Kent State University Press, Kent.

Bates, Robert L., and Julia A. Jackson (editors)

1984 *Dictionary of Geological Terms*, third edition. Anchor Press/Doubleday, Garden City.

Bayly, M. B.

1960 Errors in Point-Counter Analysis. *The American Mineralogist* 45(3–4): 447–449.

Bhattacharyya, Gouri K., and Richard A. Johnson

1977 *Statistical Concepts and Methods.* John Wiley & Sons, New York.

Brose, David S.

1978 Comments. *Current Anthropology* 19(4):729–731.

1982 Archaeological Investigations of a Fort Ancient Community Near Ohio Brush Creek, Adams County, Ohio. *Kirtlandia* 34. The Cleveland Museum of Natural History, Cleveland.

Bullen, Ripley P.

1953 The Famous Crystal River Site. *The Florida Anthropologist* 6(1):9–37.

Butler, Brian M., and Richard W. Jefferies

1986 Crab Orchard and Early Woodland Cultures in the Middle South. In *Early Woodland Archeology*, edited by Kenneth B. Farnsworth and Thomas E. Emerson, pp. 523–534. Center for American Archeology Press, Kampsville, Illinois.

Butler, Brian M., and Robert J. Scott

2008 Early Woodland in the Lower Ohio Valley: the Southern Connection.

Paper presented at the 65th Southeastern Archaeological Conference, Charlotte, North Carolina.

Caldwell, Joseph R.

1964 Interaction Spheres in Prehistory. In *Hopewellian Studies*, edited by Joseph R. Caldwell and Robert L. Hall, pp. 133–143. Illinois State Museum Scientific Papers, Volume 12. Springfield.

Carr, Christopher

2005 Rethinking Interregional Hopewellian "Interaction." In *Gathering Hopewell*, edited by Christopher Carr and Troy Case, pp. 575–623. Kluwer Academic/Plenum Publishers, New York.

2008 Settlement and Communities. In *The Scioto Hopewell and Their Neighbors*, edited by D. Troy Case and Christopher Carr, pp. 101–150. Springer, New York.

Chapman, Jefferson

1973 *The Icehouse Bottom Site, 40MR23*. Report of Investigations No. 13, Department of Anthropology, University of Tennessee, Knoxville.

Chapman, Jefferson, and Bennie C. Keel

1979 Candy Creek-Connestee Components in Eastern Tennessee and Western North Carolina and Their Relationship With Adena-Hopewell. In *Hopewell Archaeology, The Chillicothe Conference*, edited by David S. Brose and N'omi Greber, pp. 157–161. The Kent State University Press, Kent.

Chayes, Felix

1956 *Petrographic Modal Analysis*. John Wiley & Sons, Inc. New York.

Cochran, W. G.

1963 *Sampling Techniques*. 2nd edition. John Wiley & Sons, New York.

Coughlin, Sean, and Mark F. Seeman

1997 Hopewellian Settlements at the Liberty Earthworks, Ross County, Ohio. In *Ohio Hopewell Community Organization*, edited by William S. Dancey and Paul J. Pacheco, pp. 231–250. The Kent State University Press, Kent.

Dancey, William S.

1991 Middle Woodland Settlement in Central Ohio: A Preliminary Report on the Murphy Site (33LI212). *Pennsylvania Archaeologist* 61(2):37–72.

Felumlee, Gary

1983 Report on the Knight Hollow Rockshelter. *Ohio Archaeologist* 33(4):22–28.

Fenneman, Nevin M.

1938 *Physiography of Eastern United States*. McGraw-Hill Book Company, Inc. New York.

Fenton, James P., and Richard W. Jefferies

1991 The Camargo Mound and Earthworks: Preliminary Findings. In *The Human Landscape in Kentucky's Past*, edited by Charles Stout and Christine K. Hensley, 40–55. Kentucky Heritage Council, Frankfort.

Floyd, Robert J.

1965 *Tennessee Rock and Mineral Resources*. State of Tennessee, Department of Conservation, Division of Geology, Bulletin 66. Nashville.

Fowler, Melvin L.
1957 *Rutherford Mound, Hardin County, Illinois*. Illinois State Museum Scientific Papers 7(1). Springfield.

Greber, N'omi
1983 *Recent Excavations at the Edwin Harness Mound, Liberty Works, Ross County, Ohio.* MCJA Special Paper No. 5. The Kent State University Press, Kent.

Griffin, James B.
1945 The Ceramic Affiliations of the Ohio Valley Adena Culture. In *The Adena People*, edited by W. S. Webb and C. E. Snow. University of Kentucky, Reports in Anthropology and Archaeology 6:220–246.

Hansen, Michael C.
1987 The Teays River. *Ohio Geology Newsletter*, Summer 1987. Division of Geological Survey, Columbus.

Henderson, A. Gwen, and David Pollach
1985 The Late Woodland Occupation of the Bentley Site. In *Woodland Period Research in Kentucky*, edited by David Pollack, Thomas N. Sanders, and Charles D. Hockensmith, pp. 140–164. Kentucky Heritage Council, Frankfort.

Heye, George G.
1919 Certain Mounds in Haywood County, North Carolina. *Contributions from the Museum of the American-Heye Foundation* 2. New York.

Jefferies, Richard W.
1976 *The Tunacunnhee Site: Evidence of Hopewell Interaction in Northwest Georgia.* Anthropological Papers of the University Georgia, Number 1. The University of Georgia, Athens.
1979 The Tunacunnhee Site: Hopewell in Northwest Georgia. In *Hopewell Archaeology, The Chillicothe Conference*, edited by David S. Brose and N'omi Greber, pp. 162–170. The Kent State University Press, Kent.
2006 Death Rituals at the Tunacunnhee Site. In *Recreating Hopewell*, edited by Douglas K. Charles and Jane E. Buikstra, pp. 161–177. University Press of Florida, Gainesville.

Judd, Neil M.
1954 *The Material Culture of Pueblo Bonito*. Smithsonian Miscellaneous Collections, vol. 124. Washington, D.C.

Keel, Bennie C.
1976 *Cherokee Archaeology*. The University of Tennessee Press, Knoxville.

Keith, Scot J.
2010a *Archaeological Data Recovery at the Leake Site, Bartow County, Georgia.* Final Report, Two Volumes. Contract report submitted to the Georgia Department of Transportation. Southern Research, Historic Preservation Consultants, Ellerslie, Georgia. (Vol. 1 online at core.tdar.org/document 376155).
2010b The Mann Site and the Leake Site: Linking the Midwest and the Southeast During the Middle Woodland Period. Paper presented at the Mid-

west Archaeological Conference in Bloomington, Indiana. (Available on line at: http://tinyurl.com/n9rtk8z.)

Kellar, James H.

1979 The Mann Site and "Hopewell" in the Lower Wabash-Ohio Valley. In *Hopewell Archaeology, The Chillicothe Conference*, edited by David S. Brose and N'omi Greber, pp. 100–107. The Kent State University Press, Kent.

Kellar, James H., A. R. Kelly, and E. V. McMichael

1962 The Mandeville Site in Southwest Georgia. *American Antiquity* 27(3): 336–355.

Lamborn, Raymond E., Chester R. Austin, and Downs Schaaf

1938 *Shales and Surface Clays of Ohio*. Fourth Series, Bulletin 39, Geological Survey of Ohio., Columbus.

Le Maitre, R. W. (editor)

2002 *Igneous Rocks, A Classification and Glossary of Terms*. Cambridge University Press, Cambridge.

Lepper, Bradley T.

1988 A Historical Review of Archaeological Research at the Newark Earthworks. *Journal of the Steward Anthropological Society* 18:118–140.

Lepper, Bradley T., and Richard W. Yerkes

1997 Hopewellian Occupations at the Northern Periphery of the Newark Earthworks: The Newark Expressway Sites Revisited. In *Ohio Hopewell Community Organization*, edited by William S. Dancey and Paul J. Pacheco, pp. 175–205. The Kent State University Press, Kent.

Maggetti, Marino

1982 Phase Analysis and Its Significance for Technology and Origin, In *Archaeological Ceramics*, edited by G.S. Olin and A.D. Franklin, pp. 121–133. Smithsonian Institution Press, Washington, D.C.

Mainfort, Robert C., Jr.

1986 *Pinson Mounds, A Middle Woodland Ceremonial Center*. Research Series No. 7, Tennessee Department of Conservation, Division of Archaeology. Nashville.

2013 *Pinson Mounds, Middle Woodland Ceremonialism in the Midsouth*. The University of Arkansas Press, Fayetteville.

Mainfort, Robert C., Jr., James C. Cogswell, Michael J. O'Brien, Hector Neff, and Michael D. Glascock

1997 Neutron Activation Analysis of Pottery from Pinson Mounds and Nearby Sites in Western Kentucky *(sic.)*: Local Production vs. Long-Distance Importation. *Midcontinental Journal of Archaeology* 22(1):43–68.

Milanich, Jerald T.

1994 *Archaeology of Precolumbian Florida*. University of Florida Press, Gainesville.

Milanich, Jerald T., and Charles H. Fairbanks

1980 *Florida Archaeology*. Academic Press, New York.

Mills, William C.
1916 Exploration of the Tremper Mound. *Ohio Archaeological and Historical Quarterly* 25:262–398. Columbus.
1921 Flint Ridge. *Ohio Archaeological and Historical Quarterly* 30:91–161. Columbus.

Moore, Clarence B.
1903 *Certain Aboriginal Mounds of the Florida Central West-Coast.* Journal of the Academy of Natural Sciences of Philadelphia 12:361–439.
1907 *Crystal River Revisited.* Journal of the Academy of Natural Sciences of Philadelphia 13:406–425.
1918 *The Northwest Florida Coast Revisited.* Journal of the Academy of Natural Sciences of Philadelphia 16:513–581.

Morgan, Richard G.
1952 Outline of Cultures in the Ohio Region. In *Archeology of Eastern United States*, edited by James B. Griffin, pp. 83–98. The University of Chicago Press, Chicago.

Neff, Hector, Jeffrey Blomster, Michael D. Glascock, Ronald Bishop, M. James Blackman, Michael D. Coe, George L. Cowgill, Richard A. Diehl, Stephen Houston, Arthur A. Joyce, Carl P. Lipo, Barbara L. Stark, and Marcus Winter
2006 Methodological Issues in the Provenience Investigation of Early Formative Mesoamerican Ceramics. *Latin American Antiquity* 17(1):54–76.

Noble, Allen G., and Albert J. Korsok
1975 *Ohio—An American Heartland.* State of Ohio, Division of Geological Survey, Columbus.

Phillips, Philip
1970 *Archaeological Survey in the Lower Yazoo Basin, Mississippi, 1949–1955.* Papers of the Peabody Museum of Archaeology and Ethnollogy, vol. 60. Harvard University, Cambridge.

Phillips, Philip, James A. Ford, and James B. Griffin
1951 *Archaeological Survey in the Lower Mississippi Alluvial Valley, 1940–1947.* Papers of the Peabody Museum of American Archaeology and Ethnology, vol. 25. Harvard University, Cambridge.

Phillips, William Revel, and Dana T. Griffen
1981 Optical Mineralogy, The Nonopaque Minerals. W. H. Freeman, San Francisco.

Pickard, William H.
1996 Excavations at Capitolium Mound (33WN13), Marietta, Washington County, Ohio: A Working Evaluation. In *A View from the Core, A Synthesis of Ohio Archaeology*, edited by Paul J. Pacheco, pp. 274–285. The Ohio Archaeological Council, Columbus.

Pluckhahn, Thomas J.
2003 *Kolomoki.* The University of Alabama Press, Tuscaloosa.

Prufer, Olaf H.
1964 The Hopewell Complex of Ohio. In *Hopewellian Studies,* edited by Joseph R. Caldwell and Robert L. Hall, pp. 35–83. Illinois State Museum Scientific Papers, vol. 12. Springfield.
1965 *The McGraw Site, A Study in Hopewellian Dynamics*. Scientific Publications of the Cleveland Museum of Natural History, New Series, vol. 4, No. 1. Cleveland.
1968 *Ohio Hopewell Ceramics, An Analysis of the Extant Collections.* Anthropological Papers, Museum of Anthropology, University of Michigan, No. 33. Ann Arbor.

Prufer, Olaf H., and Douglas H. McKenzie
1965 Ceramics. In *The McGraw Site, A Study in Hopewellian Dynamics,* edited by Olaf H. Prufer, pp.18–59. Scientific Publications of the Cleveland Museum of Natural History, New Series, vol. 4, No. 1. Cleveland.

Richmond, Michael D., and Jonathan P. Kerr
2005 Middle Woodland Ritualism in the Central Bluegrass. In *Woodland Period Systematics in the Middle Ohio Valley*, edited by Darlene Applegate and Robert C. Mainfort Jr., pp. 76–93. University of Alabama Press, Tuscaloosa.

Rohm, Mackenzie Smyth Caldwell
2008 A Terminal Middle Woodland Ceramic Complex from Southern Illinois. Unpublished Master's Thesis, Southern Illinois University Carbondale.

Ruby, Brett J.
1997 The Mann Phase: Hopewellian Subsistence and Settlement Adaptations in the Wabash Lowlands of Southwestern Indiana. Ph.D. dissertation, Department of Anthropology, University of Indiana, Bloomington.
2006 The Mann Phase. In *Recreating Hopewell*, edited by Douglas K. Charles and Jane E. Buikstra, pp. 190–205. University Press of Florida, Gainesville.

Ruby, Bret J., Christopher Carr, and Douglas K. Charles
2005 Community Organizations in the Scioto, Mann, and Havana Hopewellian Regions. In *Gathering Hopewell*, edited by Christopher Carr and D. Troy Case, pp. 119–176. Kluwer Academic/Plenum Publishers, New York.

Ruby, Bret J., and Christine M. Shriner
2005 Ceramic Vessel Compositions and Styles as Evidence of the Local and Nonlocal Social Affiliations of Ritual Participants at the Mann Site, Indiana. In *Gathering Hopewell*, edited by Christopher Carr and D. Troy Case, pp. 553–572. Kluwer Academic/Plenum Publishers, New York.

Schubert, P.
1986 Petrographic Modal Analysis—A Necessary Complement to Chemical Analysis of Ceramic Coarse Ware. *Archaeometry*, 28(2)163–178.

Sears, William H.
1956 *Excavations at Kolomoki: Final Report.* University of Georgia Press, Athens.
1962 Hopewellian Affiliations of Certain Sites on the Gulf Coast of Florida. *American Antiquity* 28(1):5–18.

Seeman, Mark F.
1979 *The Hopewell Interaction Sphere: The Evidence for Interregional Trade and Structural Complexity.* Indiana Historical Society, Prehistory Research Series 5(2):235–438. Indianapolis.

Seeman, Mark F., and Frank Soday
1980 The Russell Brown Mounds: Three Hopewell Mounds in Ross County, Ohio. *Midcontinental Journal of Archaeology* 5(1):73–116.

Shepard, Anna O.
1936 The Technology of Pecos Pottery. In *The Pottery of Pecos, vol. II*, edited by A.V. Kidder, pp. 389–588. Yale University Press, New Haven.
1939 Appendix A—Technology of La Plata Pottery. In *Archaeological Studies in the La Plata District*, by Earl H. Morris, pp. 249–287. Carnegie Institution of Washington, Publication 519, Washington, D.C.
1954 Letter excerpt, In *The Material Culture of Pueblo Bonito*, by Neil M. Judd, pp. 236–238. Smithsonian Miscellaneous Collections, vol. 124. Washington, D.C.
1956 *Ceramics for the Archaeologist.* Publication 609. Carnegie Institution of Washington, Publication 519, Washington, D.C.

Smith, Betty A.
1975 The Relationship Between Deptford and Swift Creek Ceramics as Evidenced at the Mandeville Site, 9 Cla 1. *Southeastern Archaeological Conference, Bulletin* 18:195–200.
1979 The Hopewell Connection in Southwest Georgia In *Hopewell Archaeology, The Chillicothe Conference*, edited by David S. Brose and N'omi Greber, pp. 181–187. The Kent State University Press, Kent.

Squier, Ephraim G., and Edwin H. Davis
1848 *Ancient Monuments of the Mississippi Valley.* Smithsonian Contributions to Knowledge, vol 1. Smithsonian Institution, Washington, D.C.

Steinen, Karl T.
2006 Kolomoki. In *Recreating Hopewell*, edited by Douglas K. Charles and Jane E. Buikstra, pp. 178–189. University Press of Florida, Gainesville.

Steponaitis, Vincas P.
1984 Technological Studies of Prehistoric Pottery from Alabama: Physical Properties and Vessel Function. In *Many Dimensions of Pottery*, edited by Sander E. Van der Leeuw and Alison C. Pritchard, pp. 79–122. Universiteit van Amsterdam, Amsterdam.

Stoltman, James B.
1989 A Quantitative Approach to the Petrographic Analysis of Ceramic Thin Sections. *American Antiquity* 54(1): 147–160.

1991 Ceramic Petrography as a Technique for Documenting Cultural Interaction: An Example From the Upper Mississippi Valley. *American Antiquity* 56(1):103–120.

1999a The Chaco-Chuska Connection: In Defense of Anna Shepard. In *Pottery and People*, edited by James M. Skibo and Gary M. Feinman, pp. 1–24. The University of Utah Press, Salt Lake City.

1999b Icehouse Bottom and the Hopewell Connection. *Frank H. McClung Museum Research Notes*, No. 17. University of Tennessee, Knoxville.

2001 The Role of Petrography in the Study of Archaeological Ceramics. In *Earth Sciences and Archaeology*, edited by Paul Goldberg, Vance Holliday, and Reid Ferring, pp. 297–326. Kluwer Academic/Plenum Publishers, New York.

2010 Petrographic Observations on Middle Woodland Pottery from the Leake Site. In *Archaeological Data Recovery at the Leake Site, Bartow County, Georgia*, edited by Scot J. Keith. Volume II, Appendices. Appendix 1. Contract report submitted to the Georgia Department of Transportation. Southern Research, Historic Preservation Consultants, Ellerslie, Georgia.

Stoltman, James B., and Robert C. Mainfort, Jr.

2002 Minerals and Elements: Using Petrography to Reconsider the Findings of Neutron Activation in the Compositional Analysis of Ceramics from Pinson Mounds, Tennessee. *Midcontinental Journal of Archaeology* 27(1):1–33.

Stoltman, James B., and Frankie Snow

2002 Cultural Interaction Within Swift Creek Society: People, Pots, and Paddles. In *A World Engraved, Archaeology of the Swift Creek Culture*, edited by Mark Williams and Daniel T. Elliot, pp. 130–153. The University of Alabama Press, Tuscaloosa.

Stout, Wilber

1916 *Geology of Southern Ohio*. Bulletin 20, Fourth Series. Geological Survey of Ohio, Columbus.

Stout, Wilbur, and Downs Schlaaf

1931 Minford Silts of Southern Ohio. *Bulletin of the Geological Society of America*, 42:673–696.

Ver Steeg, Karl

1946 The Teays River. *The Ohio Journal of Science* 46(6):297–307.

Walthall, John A.

1980 *Prehistoric Indians of the Southeast, Archaeology of Alabama and the Middle South*. The University of Alabama Press, Tuscaloosa.

Weisman, Brent R.

1995 Crystal River, A Ceremonial Mound Center on the Florida Gulf Coast. *Florida Archaeology* 8.

Willey, Gordon R.

1949 *Archeology of the Florida Gulf Coast*. Smithsonian Miscellaneous Collections, Volume 113. The Smithsonian Institution, Washington, D.C.

Williams, D. F.
1983 Petrology of Ceramics. In *The Petrology of Archaeological Artefacts*, edited by D. R. C. Kemps and Anthony P. Harvey, pp. 301–329. Clarendon Press, Oxford.

Williams, Mark, and Daniel T. Elliot (editors)
1998 *A World Engraved, Archaeology of the Swift Creek Culture*. The University of Alabama Press, Tuscaloosa.

Willoughby, C. C., and E. A. Hooten
1922 *The Turner Group of Earthworks, Hamilton County, Ohio.* Harvard University, Peabody Museum of American Archaeology and Ethnology, Papers 8(3). Cambridge.

Index